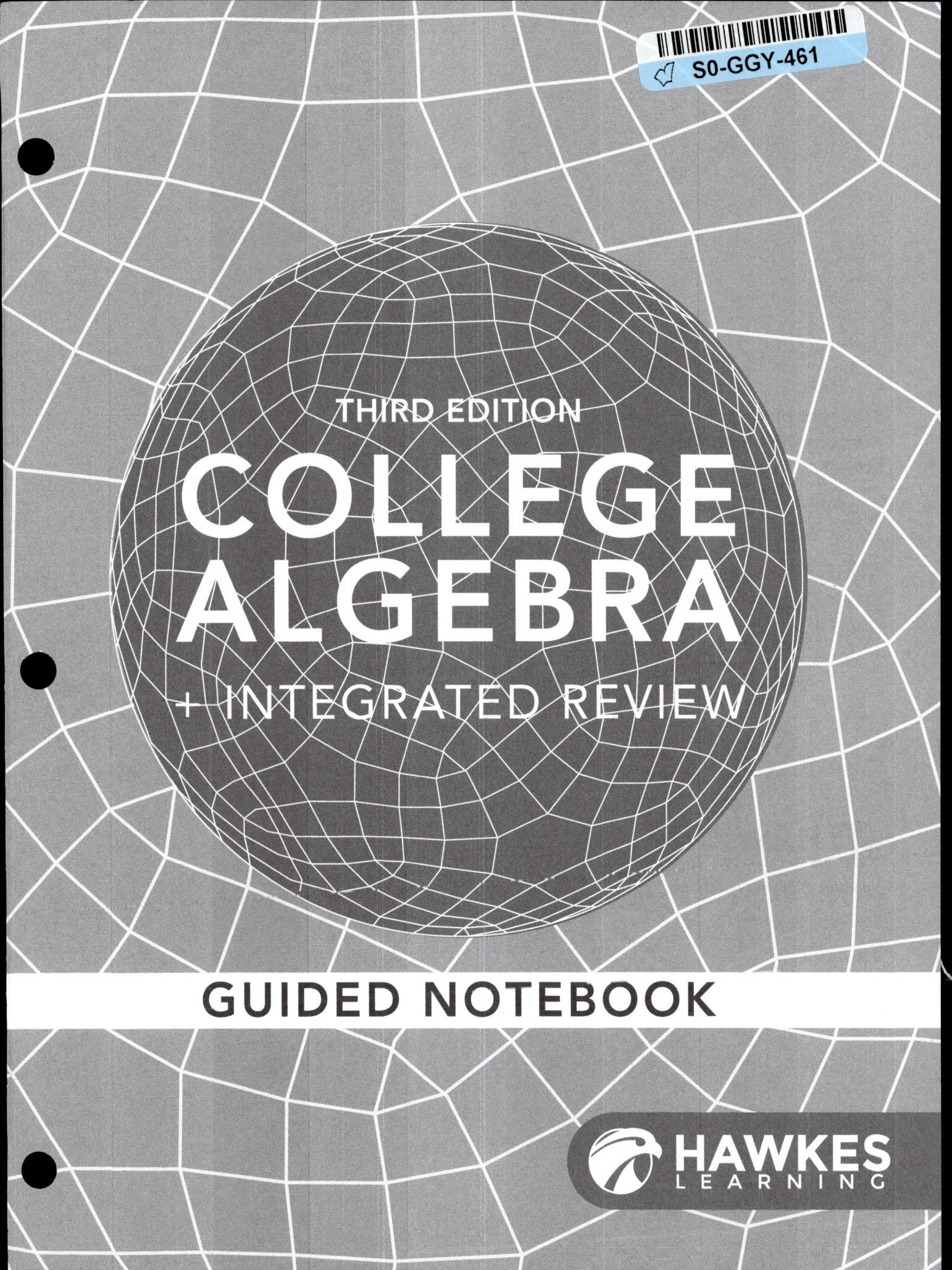

Manager of Math Content Development:
Chelsey Cooke

Editor:
Daniel Breuer

Assistant Editors:
Allison Conger,
Douglas Chappell

Creative Services Manager:
Trudy Gove

Designer:
Joel Travis

Cover Design:
Patrick Thompson

Content Contributor:
Barbara Miller

Composition and Answer Key Assistance:
QSI (Pvt.) Ltd.

A division of Quant Systems, Inc.

546 Long Point Road
Mount Pleasant, SC 29464

Copyright © 2020 by Hawkes Learning / Quant Systems, Inc. All rights reserved.

No part of this publication may be reproduced, stored in a retrieval system, or transmitted in any form or by any means, electronic, mechanical, photocopying, recording, or otherwise, without the prior written consent of the publisher.

Printed in the United States of America

10 9 8 7 6 5 4 3 2

ISBN: 978-1-64277-304-0

Table of Contents

Preface
Math Knowledge Required for Math@Work Career Explorations vi
How to Use the Guided Notebook ix

Strategies for Academic Success
How to Read a Math Textbook xiv
Tips for Success in a Math Course xv
Tips for Improving Math Test Scores xvi
Practice, Patience, and Persistence! xvii
Note Taking xviii
Do I Need a Math Tutor? xix
Tips for Improving Your Memory xx
Overcoming Anxiety xxi
Online Resources xxii
Preparing for a Final Math Exam xxiii
Managing Your Time Effectively xxv

CHAPTER 1.R
Review Concepts
for *Fundamental Concepts of Algebra*
1.R.1 Exponents, Prime Numbers, and LCM 3
1.R.2 Multiplication and Division with Fractions 19
1.R.3 Addition and Subtraction with Fractions 27
1.R.4 Proportions 33
1.R.5 Decimals, Fractions, and Percents 39

CHAPTER 2.R
Review Concepts
for *Equations and Inequalities in One Variable*
2.R.1 The Real Number Line and Absolute Value 49
2.R.2 Addition with Real Numbers 55
2.R.3 Subtraction with Real Numbers 61
2.R.4 Multiplication and Division with Real Numbers 65

CHAPTER 3.R

Review Concepts
for *Equations and Inequalities in Two Variables*

3.R.1 Formulas in Geometry	73
3.R.2 Square Roots and the Pythagorean Theorem	89
3.R.3 Evaluating Radicals	95
3.R.4 Simplifying Radicals	101
3.R.5 Introduction to the Cartesian Coordinate System	107
3.R.6 Solving Linear Equations: $ax + b = c$	115
3.R.7 Solving Linear Equations: $ax + b = cx + d$	121
3.R.8 Solving Linear Inequalities in One Variable	127
3.R.9 Solving Radical Equations	135

CHAPTER 4.R

Review Concepts
for *Relations, Functions, and Their Graphs*

4.R.1 Introduction to Functions and Function Notation	143
4.R.2 Translating English Phrases and Algebraic Expressions	151
4.R.3 Applications: Number Problems and Consecutive Integers	157
4.R.4 Greatest Common Factor (GCF) and Factoring by Grouping	163
4.R.5 Factoring Trinomials: $x^2 + bx + c$	169
4.R.6 Factoring Trinomials: $ax^2 + bx + c$	175
4.R.7 Review of Factoring Techniques	181
4.R.8 Solving Quadratic Equations by Factoring	185
4.R.9 Multiplication and Division with Complex Numbers	191
4.R.10 Quadratic Equations: The Quadratic Formula	197

CHAPTER 5.R

Review Concepts
for *Working with Functions*

5.R.1 Order of Operations with Real Numbers	205
5.R.2 Simplifying and Evaluating Algebraic Expressions	211
5.R.3 Multiplication with Polynomials	217
5.R.4 Division with Polynomials	223
5.R.5 Introduction to Rational Expressions	229
5.R.6 Multiplication and Division with Rational Expressions	237
5.R.7 Simplifying Complex Fractions	243

CHAPTER 7.R

Review Concepts
for *Exponential and Logarithmic Functions*

7.R.1 Rules for Exponents 251
7.R.2 Power Rules for Exponents 257
7.R.3 Rational Exponents 261
7.R.4 Introduction to Logarithmic Functions 267

CHAPTER 8.R

Review Concepts
for *Conic Sections*

8.R.1 Special Products of Binomials 275
8.R.2 Special Factoring Techniques 283

CHAPTER 9.R

Review Concepts
for *Systems of Equations and Inequalities*

9.R.1 Systems of Linear Equations: Solutions by Graphing 291
9.R.2 Systems of Linear Equations: Solutions by Substitution ... 299
9.R.3 Systems of Linear Equations: Solutions by Addition 305
9.R.4 Systems of Linear Inequalities 311

Math@Work

Basic Inventory Management 319
Hospitality Management: Preparing for a Dinner Service 321
Bookkeeper 323
Pediatric Nurse 325
Architecture 327
Statistician: Quality Control 329
Dental Assistant 331
Financial Advisor 333
Market Research Analyst 335
Chemistry 337
Astronomy 339
Math Education 341
Physics 343
Forensic Scientist 345
Other Careers in Mathematics 347

Answer Key 349

Math Knowledge Required for Math@Work Career Explorations

The following table summarizes the math knowledge required for each Math@Work career exploration. Use this table to determine when you are ready to explore each career.

Math@Work Career	Whole Numbers	Fractions	Integers	Decimal Numbers	Averages	Percents	Simple Interest	Ratios	Proportions	Geometry	Statistics	Graphing	Linear Equations	Systems of Equations	Mixture Problems	Scientific Notation	Greatest Common Factor	Rational Expressions	Radicals
Basic Inventory Management	✓																		
Hospitality Management	✓	✓			✓														
Bookkeeper				✓															
Pediatric Nurse				✓				✓	✓										
Architecture				✓						✓									
Statistician: Quality Control				✓							✓	✓							
Dental Assistant	✓			✓		✓													
Financial Advisor				✓		✓	✓							✓					
Market Research Analyst				✓										✓					
Chemistry				✓											✓	✓			
Astronomy				✓												✓			
Math Education			✓														✓		
Physics		✓																✓	
Forensic Scientist				✓		✓													✓
Other Careers in Mathematics																			

🛟 Support

If you have questions or comments we can be contacted as follows:

24/7 Chat: chat.hawkeslearning.com

Phone: (843) 571-2825

E-mail: support@hawkeslearning.com

Web: hawkeslearning.com

Our support hours are 8:00 a.m. to 10:00 p.m. (ET), Monday through Friday.

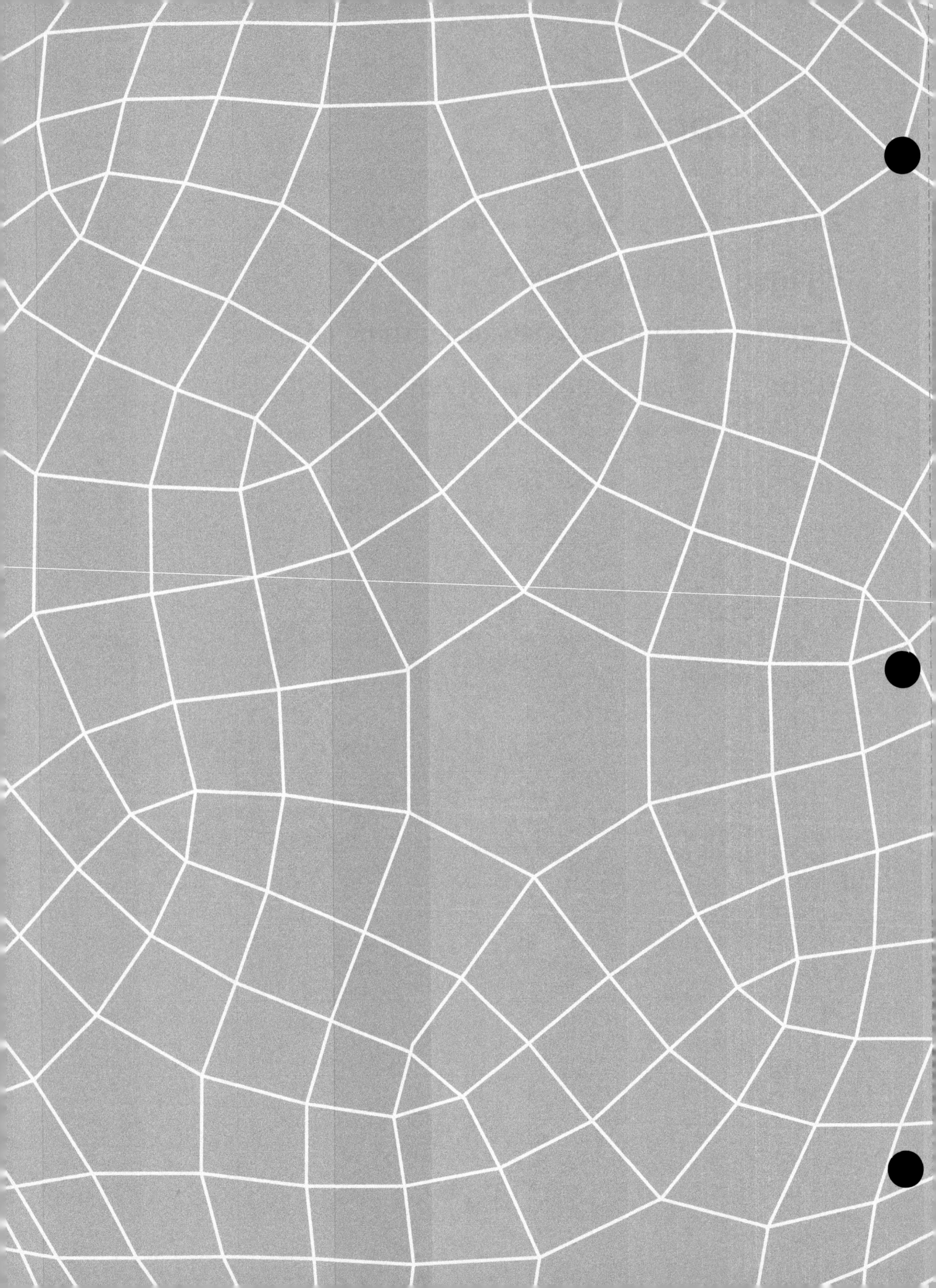

How to Use the Guided Notebook

There are a variety of elements in this Guided Notebook that will help you on your way to mastering each topic. Here is a rundown of how to use the elements as you work through this notebook.

Fill-in-the-Blanks

1. When there is an incomplete sentence, you will need to write in the __*missing*__ word(s).

   The __*missing*__ words can be found by reading through the Learn screens.

Boxed Content

Definitions and **procedures** are highlighted within a box, like the ones shown here. The missing content will vary from box to box. Sometimes an entire definition is missing and sometimes only part of a sentence is missing. Here are two examples of the box variations.

Definition

First term to define: __*Write the definition here.*__

Second term to define: __*If there is another term, define it the same way as above.*__

DEFINITION

Terms Related to Probability

__*Outcome*__	An individual result of an experiment.
__*Sample Space*__	The set of all possible outcomes of an experiment.
__*Event*__	Some (or all) of the outcomes from the sample space.

DEFINITION

How to Use the Guided Notebook

Properties and **Procedure** boxes are completed in a similar way:

Commutative Property of Multiplication

The order of the numbers in multiplication can be __reversed without changing the product.__

For example, __3 · 4 = 12 and 4 · 3 = 12.__

PROPERTIES

Subtracting Whole Numbers

1. Write the numbers __vertically__ so that the __place values are lined up in columns.__
2. Subtract only the __digits with the same place value.__
3. Check by __adding the difference to the subtrahend.__ The sum must be __the minuend.__

PROCEDURE

▶ Watch and Work

For each Watch and Work, you will need to watch the corresponding video in Learn mode and follow along while completing the example in the space provided.

Example 5 Multiplying Whole Numbers

Multiply: 12 · 35

Solution

The standard form of multiplication is used here to find the product 12 · 35.

$$\begin{array}{r} \overset{1}{1}2 \\ \times\ 35 \\ \hline 60 \\ 360 \\ \hline 420 \end{array}$$

12 · 5 = 60
12 · 30 = 360
Product

✏ Now You Try It!

After working along with the example video, work through a similar exercise on your own in the space provided.

Example A Multiplying Whole Numbers

Multiply: 25
          × 42
          ____
          1050

1.1 Exercises

Each section has exercises to offer additional practice problems to help reinforce topics that have been covered. The exercises include Concept Check, Practice, Application, and Writing & Thinking questions. The odd answers can be found in the Answer Key at the back of the book.

Concept Check

True/False. Determine whether each statement is true or false. If a statement is false, explain how it can be changed so the statement will be true. (**Note:** There may be more than one acceptable change.)

1. When the given statement is true, you write "True" for the answer.

    True

Practice

For each set of data, find **a.** the mean, **b.** the median, **c.** the mode (if any), and **d.** the range.

2. *Presidents:* The ages of the first five US presidents of the 20th century on the date of their inaugurations were as follows. (The presidents were Roosevelt, Taft, Wilson, Harding, and Coolidge.)

    42,   51,   56,   55,   51

    a. 51   b. 51   c. 51   d. 14

Applications

Solve.

3. *Grades:* Suppose that you have taken four exams and have one more chemistry exam to take. Each exam has a maximum of 100 points and you must average between 75 and 82 points to receive a passing grade of C. If you have scores of 85, 60, 73, and 76 on the first four exams, what is the minimum score you can make on the fifth exam and receive a grade of C?

    81

Writing & Thinking

4. State how to determine the median of a set of data.

    The first step to finding the median is always to arrange the data in order. Once the data is in order, the median is the number in the middle. If there is an even number of items, average the two middle numbers to find the median.

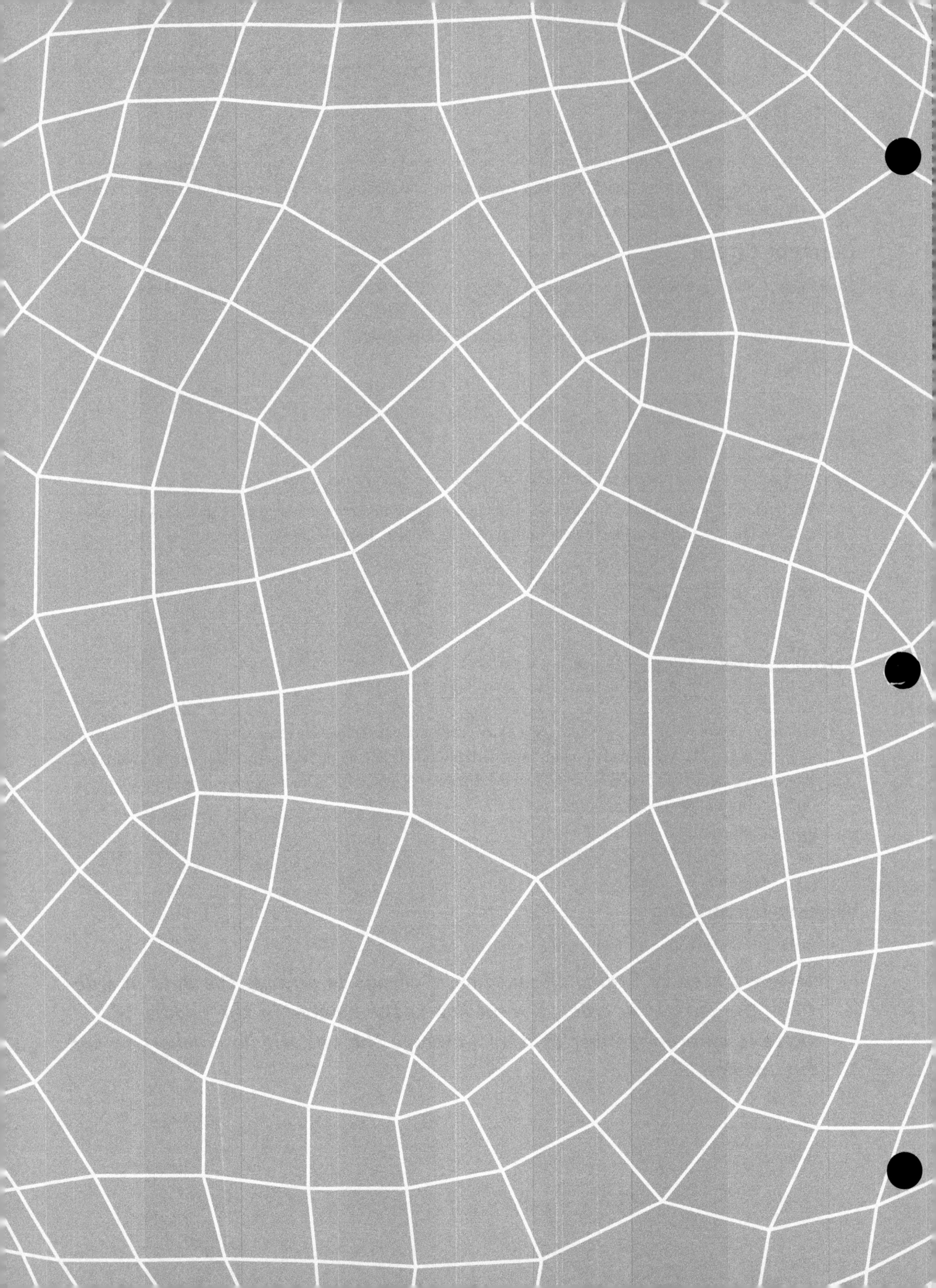

Strategies for Academic Success

Strategies for Academic Success
How to Read a Math Textbook

Reading a textbook is very different than reading a book for fun. You have to concentrate more on what you are reading because you will likely be tested on the content. Reading a math textbook requires a different approach than reading literature or history textbooks because the math textbook contains a lot of symbols and formulas in addition to words. Here are some tips to help you successfully read a math textbook.

Don't Skim

When reading math textbooks, look at everything: titles, learning objectives, definitions, formulas, text in the margins, and any text that is highlighted, outlined, or in bold. Also pay close attention to any tables, figures, charts, and graphs.

Minimize Distractions

Reading a math textbook requires much more concentration than a novel by your favorite author, so pick a study environment with few distractions and a time when you are most attentive.

Start at the Beginning

Don't start in the middle of an assigned section. Math tends to build on previously learned concepts and you may miss an important concept or formula that is crucial to understanding the rest of the material in the section.

Highlight and Annotate

Put your book to good use and don't be afraid to add comments and highlighting. If you don't understand something in the text, reread it a couple of times. If it is still not clear, note the text with a question mark or some other notation so you can ask your instructor about it.

Go through Each Step of Each Example

Make sure you understand each step of an example. If you don't understand something, mark it so you can ask about it in class. Sometimes math textbooks leave out intermediate steps to save space. Try working through the examples on your own, filling in any missing steps.

Take Notes *< This is important!*

Write down important definitions, symbols or notation, properties, formulas, theorems, and procedures. Review these daily as you do your homework and before taking quizzes and tests. Practice rewriting definitions in your own words so you understand them better.

Notes 9-25-17:
- *The opposite of a negative integer is a positive integer.*
- *To add two integers with the same signs add their absolute values and use their common sign*

Use Available Resources

Many textbooks have companion websites to help you understand the content. These resources may contain videos that help explain more complex steps or concepts. Try searching the internet for additional explanations of topics you don't understand.

Read the Material Before Class

Try to read the material from your book before the instructor lectures on it. After the lecture, reread the section again to help you retain the information as you look over your class notes.

Understand the Mathematical Definitions + × =

Many terms used in everyday English have a different meaning when used in mathematics. Some examples include equivalent, similar, average, median, and product. Two equations can be equivalent to one another without being equal. An average can be computed mathematically in several ways. It is important to note these differences in meaning in your notebook along with important definitions and formulas.

Try Reading the Material Aloud

Reading aloud makes you focus on every word in the sentence. Leaving out a word in a sentence or math problem could give it a totally different meaning, so be sure to read the text carefully and reread, if necessary.

Questions

1. Explain how taking notes can help you understand new concepts and skills while reading a math textbook.

2. Think of two more tips for reading a math textbook.

Strategies for Academic Success

Tips for Success in a Math Course

Read Your Textbook/Workbook

One of the most important skills when taking a math class is knowing how to read a math textbook. Reading a section before class and then reading it again afterwards is an important strategy for success in a math course. If you don't have time to read the entire assigned section, you can get an overview by reading the introduction or summary and looking at section objectives, headings, and vocabulary terms.

Take Notes

Take notes in class using a method that works for you. There are many different note-taking strategies, such as the Cornell Method and Concept Mapping. You can try researching these and other methods to see if they might work better than your current note-taking system.

Review

While the information is fresh in your mind, read through your notes as soon as possible after class to make sure they are readable, write down any questions you have, and fill in any gaps. Mark any information that is incomplete so that you can get it from the textbook or your instructor later.

Stay Organized

As you review your notes each day, be sure to label them using categories such as definition, theorem, formula, example, and procedure. Try highlighting each category with a different colored highlighter.

Use Study Aids

Use note cards to help you remember definitions, theorems, formulas, or procedures. Use the front of the card for the vocabulary term, theorem name, formula name, or procedure description. Write the definition, the theorem, the formula, or the procedure on the back of the card, along with a description in your own words.

Practice, Practice, Practice!

Math is like playing a sport. You can't improve your basketball skills if you don't practice—the same is true of math. Math can't be learned by only watching your instructor work through problems; you have to be actively involved in doing the math yourself. Work through the examples in the book, do some practice exercises at the end of the section or chapter, and keep up with homework assignments on a daily basis.

Do Your Homework

When doing homework, always allow plenty of time to finish it before it is due. Check your answers when possible to make sure they are correct. With word or application problems, always review your answer to see if it appears reasonable. Use the estimation techniques that you have learned to determine if your answer makes sense.

Understand, Don't Memorize

Don't try to memorize formulas or theorems without understanding them. Try describing or explaining them in your own words or look for patterns in formulas so you don't have to memorize them. For example, you don't need to memorize every perimeter formula if you understand that perimeter is equal to the sum of the lengths of the sides of the figure.

Study

Plan to study two to three hours outside of class for every hour spent in class. If math is your most difficult subject, then study while you are alert and fresh. Pick a study time when you will have the least interruptions or distractions so that you can concentrate.

Manage Your Time

Don't spend more than 10 to 15 minutes working on a single problem. If you can't figure out the answer, put it aside and work on another one. You may learn something from the next problem that will help you with the one you couldn't do. Mark the problems that you skip so that you can ask your instructor about it during the next class. It may also help to work a similar, but perhaps easier, problem.

> **Questions**
> 1. Based on your schedule, what are the best times and places for you to study for this class?
> 2. Describe your method of taking notes. List two ways to improve your method.

Strategies for Academic Success 🎓
Tips for Improving Math Test Scores

Preparing for a Math Test

- Avoid cramming right before the test and don't wait until the night before to study. Review your notes and note cards every day in preparation for quizzes and tests.
- If the textbook has a chapter review or practice test after each chapter, work through the problems as practice for the test.
- If the textbook has accompanying software with review problems or practice tests, use it for review.
- Review and rework homework problems, especially the ones that you found difficult.
- If you are having trouble understanding certain concepts or solving any types of problems, schedule a meeting with your instructor or arrange for a tutoring session (if your college offers a tutoring service) well in advance of the next test.

Test-Taking Strategies

- Scan the test as soon as you get it to determine the number of questions, their levels of difficulty, and their point values so you can adequately gauge how much time you will have to spend on each question.
- Start with the questions that seem easiest or that you know how to work immediately. If there are problems with large point values, work them next since they count for a larger portion of your grade.
- Show all steps in your math work. This will make it quicker to check your answers later once you are finished since you will not have to work through all the steps again.
- If you are having difficulty remembering how to work a problem, skip it and come back to it later so that you don't spend all of your time on one problem.

After the Test

- The material learned in most math courses is cumulative, which means any concepts you miss on each test may be needed to understand concepts in future chapters. That's why it is extremely important to review your returned tests and correct any misunderstandings that may hinder your performance on future tests.
- Be sure to correct any work you did wrong on the test so that you know the correct way to do the problem in the future. If you are not sure what you did wrong, get help from a peer who scored well on the test or schedule time with your instructor to go over the test.
- Analyze the test questions to determine if the majority came from your class notes, homework problems, or the textbook. This will give you a better idea of how to spend your time studying for the next test.
- Analyze the errors you made on the test. Were they careless mistakes? Did you run out of time? Did you not understand the material well enough? Were you unsure of which method to use?
- Based on your analysis, determine what you should do differently before the next test and where you should focus your time.

> **Questions**
> 1. Determine the resources that are available to you to help you prepare for tests, such as instructor office hours, tutoring center hours, and study groups.
> 2. Discuss two additional test taking strategies.

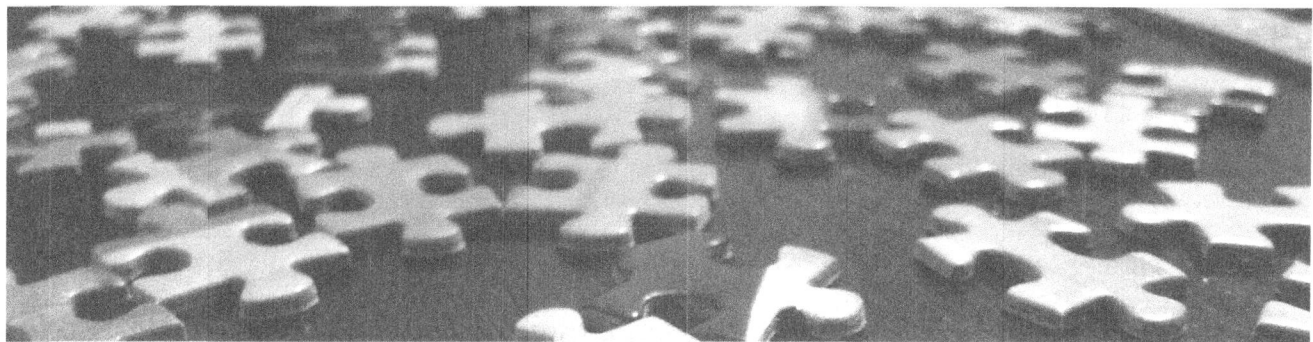

Strategies for Academic Success 🎓

Practice, Patience, and Persistence!

Have you ever heard the phrase "practice makes perfect"? This saying applies to many things in life. You won't become a concert pianist without many hours of practice. You won't become an NBA basketball star by sitting around and watching basketball on TV. The saying even applies to riding a bike. You can watch all of the videos and read all of the books on riding a bike, but you won't learn how to ride a bike without actually getting on the bike and trying to do it yourself. The same idea applies to math. Math is not a spectator sport.

Math is not learned by sleeping with your math book under your pillow at night and hoping for osmosis (a scientific term implying that math knowledge would move from a place of higher concentration—the math book—to a place of lower concentration—your brain). You also don't learn math by watching your professor do hundreds of math problems while you sit and watch. Math is learned by doing. Not just by doing one or two problems, but by doing many problems. Math is just like a sport in this sense. You become good at it by doing it, not by watching others do it. You can also think of learning math like learning to dance. A famous ballerina doesn't take a dance class or two and then end up dancing the lead in The Nutcracker. It takes years of practice, patience, and persistence to get that part.

Now, we aren't suggesting that you dedicate your life to doing math, but at this point in your education, you've already spent quite a few years studying the subject. You will continue to do math throughout college—and your life. To be able to financially support yourself and your family, you will have to find a job, earn a salary, and invest your money—all of which require some ability to do math. You may not think so right now, but math is one of the more useful subjects you will study.

It's important not only to practice math when taking a math course, but also to be patient and not expect immediate success. Just like a ballerina or NBA basketball star, who didn't become exceptional athletes overnight, it will take some time and patience to develop your math skills. Sure, you will make some mistakes along the way, but learn from those mistakes and move on.

Practice, patience, and persistence are especially important when working through applications or word problems. Most students don't like word problems and, therefore, avoid them. You won't become good at working word problems unless you practice them over and over again. You'll need to be patient when working through word problems in math since they will require more time to work than typical math skills exercises. The process of solving word problems is not a quick one and will take patience and persistence on your part to be successful.

Just as you work your body through physical exercise, you have to work your brain through mental exercise. Math is an excellent subject to provide the mental exercise needed to stimulate your brain. Your brain is flexible and it continues to grow throughout your life span—but only if provided the right stimuli. Studying mathematics and persistently working through tough math problems is one way to promote increased brain function. So, when doing mathematics, remember the 3 P's—Practice, Patience, and Persistence—and the positive effects they will have on your brain!

> **Questions**
> 1. What is another area (not mentioned here) that requires practice, patience, and persistence to master? Can you think of anything you could master without practice?
> 2. Can you think of an example in your study of math where practice, patience, and persistence have helped you improve?

Strategies for Academic Success 🎓
Note Taking

Taking notes in class is an important step in understanding new material. While there are several methods for taking notes, every note-taking method can benefit from these general tips.

General Tips

- Write the date and the course name at the top of each page.
- Write the notes in your own words and paraphrase.
- Use abbreviations, such as ft for foot, # for number, def for definition, and RHS for right-hand side.
- Copy all figures or examples that are presented during the lecture.
- Review and rewrite your notes after class. Do this on the same day, if possible.

There are many different methods of note taking and it's always good to explore new methods. A good time to try out new note-taking methods is when you rewrite your class notes. Be sure to try each new method a few times before deciding which works best for you. Presented here are three note-taking methods you can try out. You may even find that a blend of several methods works best for you.

Note-Taking Methods

Outline

An outline consists of several topic headings, each followed by a series of indented bullet points that include subtopics, definitions, examples, and other details.

> Example:
> 1. Ratio
>   a. Comparison of two quantities by division.
>   b. Ratio of a to b
>   i. $\dfrac{a}{b}$
>   ii. $a:b$
>   iii. a to b
>   c. Can be reduced
>   d. Common units can cancel

Split Page

The split page method divides the page vertically into two columns with the left column narrower than the right column. Main topics go in the left column and detailed comments go in the right column. The bottom of the page is reserved for a short summary of the material covered.

> Example:
>
> | Keywords: | Notes: |
> |---|---|
> | Ratios | 1. Comparison of two quantities by division |
> | | 2. $\dfrac{a}{b}$, $a:b$, a to b |
> | | 3. Can reduce |
> | | 4. Common units can cancel |
>
> Summary: Ratios are used to compare quantities and units can cancel.

Mapping

The mapping method is the most visual of the three methods. One common way to create a mapping is to write the main idea or topic in the center and draw lines, from the main idea to smaller ideas or subtopics. Additional branches can be created from the subtopics until all of the key ideas and definitions are included. Using a different color for subtopic can help visually organize the topics.

> Example:
>
>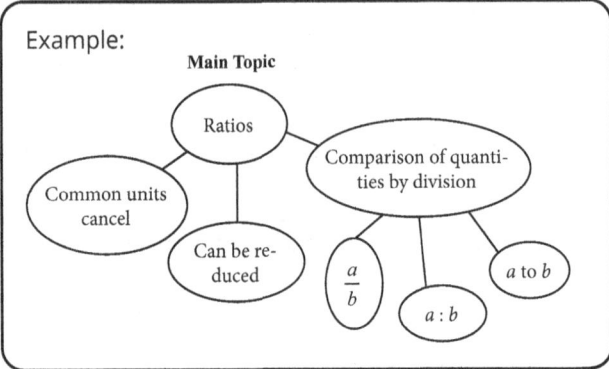

Questions

1. Find two other note taking methods and describe them.
2. Write five additional abbreviations that you could use while taking notes.

Strategies for Academic Success 🎓

Do I Need a Math Tutor?

If you do not understand the material being presented in class, if you are struggling with completing homework assignments, or if you are doing poorly on tests, then you may need to consider getting a tutor. In college, everyone needs help at some point in time. What's important is to recognize that you need help before it's too late and you end up having to retake the class.

Alternatives to Tutoring

Before getting a tutor, you might consider setting up a meeting with your instructor during their office hours to get help. Unfortunately, you may find that your instructor's office hours don't coincide with your schedule or don't provide enough time for one-on-one help.

Another alternative is to put together a study group of classmates from your math class. Working in groups and explaining your work to others can be very beneficial to your understanding of mathematics. Study groups work best if there are three to six members. Having too many people in a study group may make it difficult to schedule a time for all group members to meet. A large study group may also increase distractions. If you have too few people and those that attend are just as lost as you, then you aren't going to be helpful to each other.

Where to Find a Tutor

Many schools have both group and individual tutoring available. In most cases, the cost of this tutoring is included in tuition costs. If your college offers tutoring through a learning lab or tutoring center, then you should take advantage of it. You may need to complete an application to be considered for tutoring, so be sure to get the necessary paperwork at the start of each semester to increase your chances of getting a tutoring time that works well with your schedule. This is especially important if you know that you struggle with math or haven't taken any math classes in a while.

If you find that you need more help than the tutoring center can provide, or your school doesn't offer tutoring, you can hire a private tutor. The hourly cost to hire a private tutor varies significantly depending on the area you live in along with the education and experience level of the tutor. You might be able to find a tutor by asking your instructor for references or by asking friends who have taken higher-level math classes than you have. You can also try researching the internet for local reputable tutoring organizations in your area.

What to Look for in a Tutor

Whether you obtain a tutor through your college or hire a personal tutor, look for someone who has experience, educational qualifications, and who is friendly and easy to work with. If you find that the tutor's personality or learning style isn't similar to yours, then you should look for a different tutor that matches your style. It may take some effort to find a tutor who works well with you.

How to Prepare for a Tutoring Session

To get the most out of your tutoring session, come prepared by bringing your text, class notes, and any homework or questions you need help with. If you know ahead of time what you will be working on, communicate this to the tutor so they can also come prepared. You should attempt the homework prior to the session and write notes or questions for the tutor. Do not use the tutor to do your homework for you. The tutor will explain to you how to do the work and let you work some problems on your own while he or she observes. Ask the tutor to explain the steps aloud while working through a problem. Be sure to do the same so that the tutor can correct any mistakes in your reasoning. Take notes during your tutoring session and ask the tutor if he or she has any additional resources such as websites, videos, or handouts that may help you.

> **Questions**
> 1. It's important to find a tutor whose learning style is similar to yours. What are some ways that learning styles can be different?
> 2. What sort of tutoring services does your school offer?

Strategies for Academic Success 🎓

Tips for Improving Your Memory

Experts believe that there are three ways that we store memories: first in the sensory stage, then in short term memory, and finally in long term memory.[1] Because we can't retain all the information that bombards us daily, the different stages of memory act as a filter. Your sensory memory lasts only a fraction of a second and holds your perception of a visual image, a sound, or a touch. The sensation then moves to your short term memory, which has the limited capacity to hold about seven items for no more than 20 to 30 seconds at a time. Important information is gradually transferred to long term memory. The more the information is repeated or used, the greater the chance that it will end up in long term memory. Unlike sensory and short term memory, long term memory can store unlimited amounts of information indefinitely. Here are some tips to improve your chances of moving important information to long-term memory.

Be attentive and focused on the information.
Study in a location that is free of distractions and avoid watching TV or listening to music with lyrics while studying.

Recite information aloud.
Ask yourself questions about the material to see if you can recall important facts and details. Pretend you are teaching or explaining the material to someone else. This will help you put the information into your own words.

Associate the information with something you already know.
Think about how you can make the information personally meaningful—how does it relate to your life, your experiences, and your current knowledge? If you can link new information to memories already stored, you create "mental hooks" that help you recall the information. For example, when trying to remember the formula for slope using rise and run, remember that rise would come alphabetically before run, so rise will be in the numerator in the slope fraction and run will be in the denominator.

Use visual images like diagrams, charts, and pictures.
You can make your own pictures and diagrams to help you recall important definitions, theorems, or concepts.

Split larger pieces of information into smaller "chunks."
This is useful when remembering strings of numbers, such as social security numbers and telephone numbers. Instead of remembering a sequence of digits such as 555777213 you can break it into chunks such as 555 777 213.

Group long lists of information into categories that make sense.
For example, instead of remembering all the properties of real numbers individually, try grouping them into shorter lists by operation, such as addition and multiplication.

Use mnemonics or memory techniques to help remember important concepts and facts.
A mnemonic that is commonly used to remember the order of operations is "Please Excuse My Dear Aunt Sally," which uses the first letter of the words Parentheses, Exponents, Multiplication, Division, Addition, and Subtraction to help you remember the correct order to perform basic arithmetic calculations. To make the mnemonic more personal and possibly more memorable, make up one of your own.

Use acronyms to help remember important concepts or procedures.
An acronym is a type of mnemonic device which is a word made up by taking the first letter from each word that you want to remember and making a new word from the letters. For example, the word HOMES is often used to remember the five Great Lakes in North America where each letter in the word represents the first letter of one of the lakes: Huron, Ontario, Michigan, Erie, and Superior.

> **Questions**
> 1. Create an original mnemonic or acronym for any math topic covered so far in this course.
> 2. Explain two ways you can incorporate these tips into your study routine.

---

[1] Source: http://science.howstuffworks.com/life/inside-the-mind/human-brain/human-memory2.htm

Strategies for Academic Success 🎓

Overcoming Anxiety

People who are anxious about math are often just not good at taking math tests. If you understand the math you are learning but don't do well on math tests, you may be in the same situation. If there are other subject areas in which you also perform poorly on tests, then you may be experiencing test anxiety.

How to Reduce Math Anxiety

- Learn effective math study skills. Sit near the front of your class and take notes. Ask questions when you don't understand the material. Review your notes after class and read new material before it's covered in class. Keep up with your assignments and do a lot of practice problems.
- Don't accept negative self talk such as "I am not good at math" or "I just don't get it and never will." Maintain a positive attitude and set small math achievement goals to keep you positively moving toward bigger goals.
- Visualize yourself doing well in math, whether it's on a quiz or test, or passing a math class. Rehearse how you will feel and perform on an upcoming math test. It may also help to visualize how you will celebrate your success after doing well on the test.
- Form a math study group. Working with others may help you feel more relaxed about math in general and you may find that other people have the same fears.
- If you panic or freeze during a math test, try to work around the panic by finding something on the math test that you can do. Once you gain confidence, work through other problems you know how to do. Then, try completing the harder problems, knowing that you have a large part of the test completed already.
- If you have trouble remembering important concepts during tests, do what is called a "brain drain" and write down all the formulas and important facts that you have studied on your test or scratch paper as soon as you are given the test. Do this before you look at any questions on the test. Having this information available to you should help boost your confidence and reduce your anxiety. Doing practice brain drains while studying can help you remember the concepts when the test time comes.

How to Reduce Test Anxiety

- Be prepared. Knowing you have prepared well will make you more confident and less anxious.
- Get plenty of sleep the night before a big test and be sure to eat nutritious meals on the day of the test. It's helpful to exercise regularly and establish a set routine for test days. For example, your routine might include eating your favorite food, putting on your lucky shirt, and packing a special treat for after the test.
- Talk to your instructor about your anxiety. Your instructor may be able to make accommodations for you when taking tests that may make you feel more relaxed, such as extra time or a more calming testing place.
- Learn how to manage your anxiety by taking deep, slow breaths and thinking about places or people who make you happy and peaceful.
- When you receive a low score on a test, take time to analyze the reasons why you performed poorly. Did you prepare enough? Did you study the right material? Did you get enough rest the night before? Resolve to change those things that may have negatively affected your performance in the past before the next test.
- Learn effective test taking strategies. See the study skill on Tips for Improving Math Test Scores.

Questions

1. Describe your routine for test days. Think of two ways you can improve your routine to reduce stress and anxiety.
2. Research and describe the accommodations that your instructor or school can provide for test taking.

Strategies for Academic Success 🎓
Online Resources

With the invention of the internet, there are numerous resources available to students who need help with mathematics. Here are some quality online resources that we recommend.

HawkesTV
tv.hawkeslearning.com

If you are looking for instructional videos on a particular topic, then start with HawkesTV. There are hundreds of videos that can be found by looking under a particular math subject area such as introductory algebra, precalculus, or statistics. You can also find videos on study skills.

YouTube
www.youtube.com

You can also find math instructional videos on YouTube, but you have to search for videos by topic or key words. You may have to use various combinations of key words to find the particular topic you are looking for. Keep in mind that the quality of the videos varies considerably depending on who produces them.

Google Hangouts
hangouts.google.com

You can organize a virtual study group of up to 10 people using Google Hangouts. This is a terrific tool when schedules are hectic and it avoids everyone having to travel to a central location. In addition to video chat, the group members can share documents using Google Docs. This is a great tool for group projects!

Wolfram|Alpha
www.wolframalpha.com

Wolfram|Alpha is a computational knowledge engine developed by Wolfram Research that answers questions posed to it by computing the answer from "curated data." Typical search engines search all of the data on the Internet based on the key words given and then provide a list of documents or web pages that might contain relevant information. The data used by Wolfram|Alpha is said to be "curated" because someone has to verify its integrity before it can be added to the database, therefore ensuring that the data is of high quality. Users can submit questions and request calculations or graphs by typing their request into a text field. Wolfram|Alpha then computes the answers and related graphics from data gathered from both academic and commercial websites such as the CIA's World Factbook, the United States Geological Survey, financial data from Dow Jones, etc. Wolfram|Alpha uses the basic features of Mathematica, which is a computational toolkit designed earlier by Wolfram Research that includes computer algebra, symbol and number computation, graphics, and statistical capabilities.

> **Questions**
> 1. Describe a situation where you think Wolfram|Alpha might be more helpful than YouTube, and vice versa.
> 2. What are some pros and cons to using Google Hangouts?

Strategies for Academic Success 🎓

Preparing for a Final Math Exam

Since math concepts build on one another, a final exam in math is not one you can study for in a night or even a day or two. To pull all the concepts together for the semester, you should plan to start one or two weeks ahead of time. Being comfortable with the material is key to going into the exam with confidence and lowering your anxiety.

Before You Start Preparing for the Exam

1. What is the date, time, and location of the exam? Check your syllabus for the final exam time and location. If it's not on your syllabus, your instructor should announce this information in class.

2. Is there a time limit on the exam? If you experience test anxiety on timed tests, be sure to speak to your professor about it and see if you can receive accommodations that will help reduce your anxiety, such as extended time or an alternate testing location.

3. Will you be able to use a formula sheet, calculator, and/or scrap paper on the exam? If you are not allowed to use a formula sheet, you should write down important formulas and memorize them. Most of the time, math professors will advise you of the formulas you need to know for an exam. If you cannot use a calculator on the exam, be sure to practice doing calculations by hand when you are preparing for the exam and go back and check them using the calculator.

A Week Before the Exam

1. Decide where to study for the exam and with whom. Make sure it's a comfortable study environment with few outside distractions. If you are studying with others, make sure the group is small and that the people in the group are motivated to study and do well on the exam. Plan to have snacks and water with you for energy and to avoid having to delay studying to go get something to eat or drink. Be sure and take small breaks every hour or two to keep focused and minimize frustration.

2. Organize your class notes and any flash cards with vocabulary, formulas, and theorems. If you haven't used flash cards for vocabulary, go back through your notes and highlight the vocabulary. Create a formula sheet to use on the exam, if the professor allows. If not, then you can use the formula sheet to memorize the formulas that will be on the exam.

3. Start studying for the exam. Studying a week before the exam gives you time to ask your instructor questions as you go over the material. Don't spend a lot of time reviewing material you already know. Go over the most difficult material or material that you don't understand so you can ask questions about it. Be sure to review old exams and work through any questions you missed.

3 Days Before the Exam

1. Make yourself a practice test consisting of the problem types. Don't necessarily put the questions in the order that the professor covered them in class.

2. Ask your instructor or classmates any questions that you have about the practice test so that you have time to go back and review the material you are having difficulty with.

The Night Before the Exam

1. Make sure you have all the supplies you will need to take the exam: formula sheet and calculator, if allowed, scratch paper, plain and colored pencils, highlighter, erasers, graph paper, extra batteries, etc.

2. If you won't be allowed to use your formula sheet, review it to make sure you know all the formulas. Right before going to bed, review your notes and study materials, but do not stay up all night to "cram."

3. Go to bed early and get a good night's sleep. You will do better if you are rested and alert.

The Day of the Exam

1. Get up with plenty of time to get to your exam without rushing. Eat a good breakfast and don't drink too much caffeine, which can make you anxious.

2. Review your notes, flash cards, and formula sheet again, if you have time.

3. Get to class early so you can be organized and mentally prepared.

Checklist for the Exam

Date of the Exam: _____   Time of the Exam: _____

Location of the Exam: _____

Items to bring to the exam:

___ calculator and extra batteries   ___ pencils

___ formula sheet   ___ eraser

___ scratch paper   ___ colored pencils or highlighter

___ graph paper   ___ ruler or straightedge

Notes or other things to remember for exam day:

During the Exam

1. Put your name at the top of your exam immediately. If you are not allowed to use a formula sheet, before you even look at the exam, do what is called a "brain drain" or "data dump." Recall as much of the information on your formula sheet as you possibly can and write it either on the scratch paper or in the exam margins if scratch paper is not allowed. You have now transferred over everything on your "mental cheat sheet" to the exam to help yourself as you work through the exam.

2. Read the directions carefully as you go through the exam and make sure you have answered the questions being asked. Also, check your solutions as you go. If you do any work on scratch paper, write down the number of the problem on the paper and highlight or circle your answer. This will save you time when you review the exam. The instructor may also give you partial credit for showing your work. (Don't forget to attach your scratch work to your exam when you turn it in.)

3. Skim the questions on the exam, marking the ones you know how to do immediately. These are the problems you will do first. Also note any questions that have a higher point value. You should try to work these next or be sure to leave yourself plenty of time to do them later.

4. If you get to a problem you don't know how to do, skip it and come back after you finish all the ones you know how to do. A problem you do later may jog your memory on how to do the problem you skipped.

5. For multiple choice questions, be sure to work the problem first before looking at the answer choices. If your answer is not one of the choices, then review your math work. You can also try starting with the answer choices and working backwards to see if any of them work in the problem. If this doesn't work, see if you can eliminate any of the answer choices and make an educated guess from the remaining ones. Mark the problem to come back to later when you review the exam.

6. Once you have an answer for all the problems, review the entire exam. Try working the problems differently and comparing the results or substituting the answers into the equation to verify they are correct. Do not worry about finishing early. You are in control of your own time—and your own success!

Questions

1. Does your syllabus provide any of the information needed for the checklist?
2. Are there any tips or suggestions mentioned here that you haven't thought of before?

Strategies for Academic Success 🎓

Managing Your Time Effectively

Have you ever made it to the end of a day and wondered where all of your time went? Sometimes it feels like there aren't enough hours in the day. Managing your time is important because you can never get that time back. Once it's gone, you have to rush and cram the work into your schedule. Not only will you start feeling stressed out, but you may also find yourself turning in late or incomplete work.

Here are three strategies for managing your time more effectively.

⏱ Time Budgets

Time budgets help you find the time you need to complete necessary projects and tasks. Just like a financial budget shows you how you spend your money, a time budget shows you how you spend your time. You can then identify "wasted" time that could be used more productively.

To begin budgeting your time, assess how much time each week you spend on different types of activities, like Sleep, Meals, Work, Class, Study, Extracurricular, Exercise, Personal, Other, etc.

- What are some activities you'd like to spend more time doing in the future?
- What are some activities you should spend less time doing in the future?

Based on your answers to the questions above, create a weekly time budget. One week contains only 168 hours. If you want to spend more time on a particular activity, you'll need to find that time somewhere. Use a planner to schedule specific blocks of time for study sessions, meals, travel times, and morning/evening routines. As a general rule, you should set aside at least two hours of study time for every one hour of class time. That means that a three-credit course would require at least six hours of outside work per week.

⚖️ Breaks

When you are working on an important project or studying for a big exam, you can feel tempted to go as long as possible without taking a break. While staying focused is important, working yourself until you're mentally drained will lower the quality of your work and force you to take even more time recovering.

Just like taking breaks helps your physical body recover, it will also help your brain re-energize and refocus. During study sessions, you should plan to take a break at least once an hour. Study and work breaks should usually last around five minutes. The longer the break, the harder it is to start working again. Some courses have a built-in break during the middle of the class period. Stand up and move around, even if you don't feel tired. Even this little bit of physical movement can help you think more clearly.

📋 Avoiding Multitasking

Multitasking is working on more than one task at a time. When you have several assignments that need to be completed, you may be tempted to save time by working on two or three of them at once. While this strategy might seem like a time-saver, you will probably end up using more time than if you had done each task individually. Not only will you have to switch your focus from one task to the next, but you will also make more mistakes that will need to be corrected later. Multitasking usually ends up wasting time instead of saving it.

Instead of trying to do two things at once, schedule yourself time to work on one task at a time. To-do lists can be helpful tools for keeping yourself focused on finishing one item before moving on to another. You'll do better work and save yourself time.

Questions

1. Are there any areas in your day that are taking up too much of your time, making it hard to devote enough time to more important things?
2. Can you think of a time when multitasking has resulted in lower quality outcome in your experience?

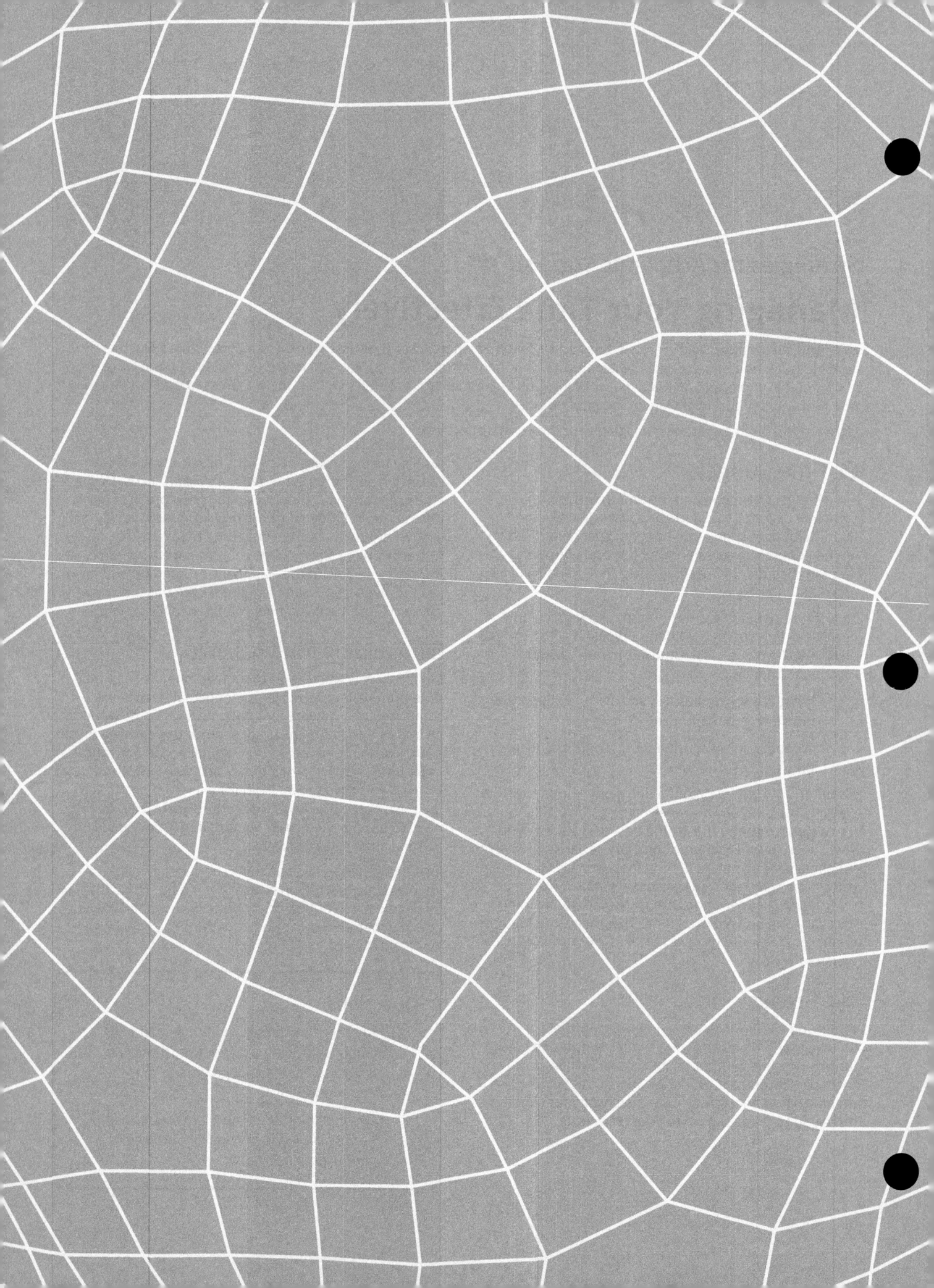

CHAPTER 1.R
Review Concepts
for *Fundamental Concepts of Algebra*

1.R.1 Exponents, Prime Numbers, and LCM

1.R.2 Multiplication and Division with Fractions

1.R.3 Addition and Subtraction with Fractions

1.R.4 Proportions

1.R.5 Decimals, Fractions, and Percents

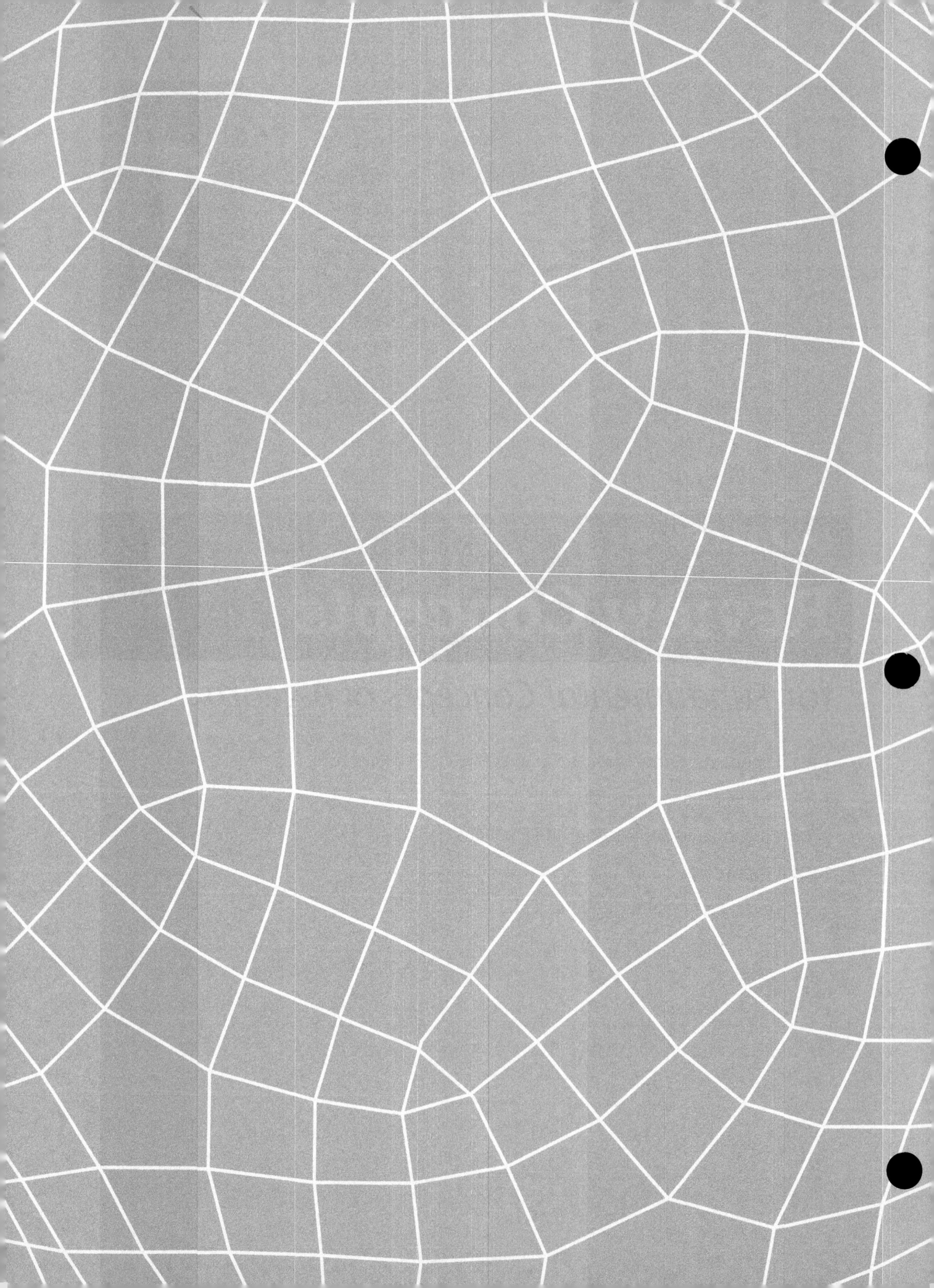

1.R.1 Exponents, Prime Numbers, and LCM

↻ Making Connections

When simplifying numeric and algebraic expressions, you will often have to work with powers of numbers, prime numbers, composite numbers, and factors of numbers. Therefore, it is important to review how to evaluate expressions with exponents, identify the prime numbers less than 50, find the prime factorization of a composite number, and recognize the LCM (least common multiple) of a set of counting numbers.

In this section, you will learn skills that you can apply when answering questions like these:

- Evaluate the following expression: $-3 + 6 \cdot 1 \div 5 + (-3)^3$
- Factor $y^2 + 2y - 15$

🛠 Building Foundations

When looking at $3^5 = 243$, 3 is the _____, 5 is the _____, and 243 is the _____. _____ are written slightly to the right and above the _____. The expression 3^5 is _____.

The Exponent 1

Any number raised to the first power _____

For example, _____

DEFINITION

The Exponent 0

Any nonzero number raised to the 0 power _____

For example, _____

Note: The expression 0^0 _____

DEFINITION

1.R.1 Exponents, Prime Numbers, and LCM

> **Rules for Order of Operations**
> 1. Simplify within grouping symbols, such as _____
>   (If there are more than one pair of grouping symbols, start with _____
>   _____
> 2. Evaluate any _____
> 3. Moving from left to right, perform any _____
>   _____
> 4. Moving from left to right, perform any _____
>   _____
>
> PROCEDURE

2. A well-known mnemonic device for remembering the rules for order of operations is the following.

Please	Excuse	My	Dear	Aunt	Sally
↓	↓	↓	↓	↓	↓
_____	_____	_____	_____	_____	_____

▶ Watch and Work

Watch the video for Example 5 in the software and follow along in the space provided.

Example 5 Using the Order of Operations with Whole Numbers

Simplify: $2 \cdot 3^2 + 18 \div 3^2$

Solution

✏️ Now You Try It!

Use the space provided to work out the solution to the next example.

Example A **Using the Order of Operations with Whole Numbers**

Simplify: $6^2 \div 9 + 3 - 14 \div 7$

Divisibility

If a number can be divided by another number so that the remainder is 0, then we say

1. the number is _____
2. the divisor _____

DEFINITION

Divisibility by 2

A number is divisible by 2 (is an **even number**) if _____

DEFINITION

Divisibility by 3

A number is divisible by 3 if _____

DEFINITION

▶ Watch and Work

Watch the video for Example 8 in the software and follow along in the space provided.

Example 8 Determining Divisibility by 3

Determine whether each of the following numbers is divisible by 3.

a. 6801

b. 356

Solution

✏ Now You Try It!

Use the space provided to work out the solution to the next example.

Example B Determining Divisibility by 3

Is 7912 divisible by 3? Explain why or why not.

1.R.1 Exponents, Prime Numbers, and LCM

Divisibility by 4
A number is divisible by 4 if _____
_____

DEFINITION

Divisibility by 5
A number is divisible by 5 if _____

DEFINITION

Divisibility by 6
A number is divisible by 6 if _____

DEFINITION

Divisibility by 9
A number is divisible by 9 if _____

DEFINITION

Divisibility by 10
A number is divisible by 10 if _____

DEFINITION

Prime Number
A **prime number** is a counting number _____
_____

DEFINITION

Composite Number
A **composite number** is _____

DEFINITION

1.R.1 Exponents, Prime Numbers, and LCM

To Determine Whether a Number is Prime

Divide the number by progressively larger prime numbers (2, 3, 5, 7, 11, and so forth) until one of the following is true.

3. The remainder _____ This means that the _____
   _____

4. You find a quotient _____ This means that the _____
   _____

PROCEDURE

The Fundamental Theorem of Arithmetic

Every composite number has _____

DEFINITION

To Find the Prime Factorization of a Composite Number

1. Factor the composite number _____

2. Factor each _____

3. Continue this process until all factors are prime.

PROCEDURE

Factors of a Composite Number

The only factors (or divisors) of a composite number are

1. _____

2. _____

3. products formed by _____

DEFINITION

▶ Watch and Work

Watch the video for Example 22 in the software and follow along in the space provided.

Example 22 Finding the Factors of a Composite Number

Find all the factors of 60.

Solution

✏ Now You Try It!

Use the space provided to work out the solution to the next example.

Example C Finding the Factors of a Composite Number

Find all the factors of 42.

1.R.1 Exponents, Prime Numbers, and LCM

The **multiples** of a number are _____

> ### Least Common Multiple (LCM)
> The **least common multiple (LCM)** of two (or more) counting numbers is _____
> _____
>
> **DEFINITION**

> ### To Find the LCM of a Set of Counting Numbers
> 1. Find the _____
> 2. List the _____
> 3. Find the product of these primes using each _____
>   _____
>
> **PROCEDURE**

▶ Watch and Work

Watch the video for Example 27 in the software and follow along in the space provided.

Example 27 Finding the Least Common Multiple (LCM)

Find the LCM of 27, 30, and 42.

Solution

✏️ Now You Try It!

Use the space provided to work out the solution to the next example.

Example D Finding the Least Common Multiple (LCM)
Find the LCM of 36, 45, and 60.

Finding Equivalent Fractions

To find a fraction equivalent to $\frac{a}{b}$, multiply the _____

_____

$$\frac{a}{b} = \frac{a}{b} \cdot \underline{\qquad\qquad}$$

For example, _____

PROCEDURE

1.R.1 Exponents, Prime Numbers, and LCM

Looking Ahead

The following example incorporates some of the skills you learned in this section, including order of operations, evaluating exponents, and finding a least common multiple in order to find a common denominator.

Example Preview

Evaluate the following expression, using the correct order of operations.

$$4 + 1 \cdot 5 \div 8 + (-1)^3$$

Solution

Because there are no grouping symbols in this expression, the first step is to calculate exponents and roots.

$$4 + 1 \cdot 5 \div 8 + (-1)^3 = 4 + 1 \cdot 5 \div 8 - 1 \quad \text{Simplify the term with an exponent.}$$

$$= 4 + \frac{5}{8} - 1 \quad \text{Perform multiplications and divisions from left to right.}$$

$$= \frac{32}{8} + \frac{5}{8} - \frac{8}{8} \quad \text{Find a common denominator.}$$

$$= \frac{29}{8} \quad \text{Add.}$$

1.R.1 Exercises

Concept Check

True/False. Determine whether each statement is true or false. If a statement is false, explain how it can be changed so the statement will be true. (**Note:** There may be more than one acceptable change.)

1. Nine squared is equal to eighteen.

2. $2^7 = 128$

3. 7^0 is undefined.

4. According to the order of operations, multiplication is always performed before division.

5. A number that is divisible by 10 is also divisible by 2 and 5.

6. 6801 is divisible by 9.

7. 7605 is divisible by 10.

8. 5,187,042 is divisible by 3.

9. A prime number has exactly 1 factor.

10. A composite number has 2 or more factors.

11. 231 is a prime number.

12. All the factors of 30 are 1, 2, 3, 5, 6, 10, 15 and 30.

13. The LCM of 15 and 25 is 50.

14. The first five multiples of 9 are 9, 18, 27, 36, and 45.

15. When given larger numbers, the most efficient way to find the LCM is to use the prime factorization method.

1.R.1 Exponents, Prime Numbers, and LCM

Practice

For each exponential expression **a.** identify the base, **b.** identify the exponent, and **c.** evaluate the exponential expression.

16. 2^3

17. 4^0

Simplify.

18. $30 \div 2 - 11 + 2(5-1)^3$

Using the tests for divisibility, determine which of 2, 3, 4, 5, 6, 9, and 10 (if any) will divide exactly into each given number.

19. 105

21. 331

22. 1234

20. 150

Determine whether each number is prime or composite. If the number is composite, find at least three factors of the number.

23. 47

24. 63

Find the prime factorization of each number. Use the tests for divisibility for 2, 3, 4, 5, 6, 9, and 10 whenever they help to find beginning factors.

25. 125

26. 150

27. Find the LCM of 3, 4, and 8.

28. For 14, 35, and 49, **a.** find the LCM and **b.** state how many times each number divides into the LCM.

For each equation, find the missing numerator that will make the fractions equivalent.

29. $\dfrac{5}{8} = \dfrac{?}{24}$

30. $\dfrac{5}{12} = \dfrac{?}{108}$

Applications

Solve.

31. **Purchases:** Robert is purchasing shirts for his weekend soccer team. The shirts he wants to buy are normally $25 each but are on sale for $10 off. His team has a total of 11 players. How much will he spend to buy the shirts?

    a. If you simplify the expression $25 − $10 · 11 using the order of operations, will you get the correct answer? If not, explain what is wrong with the expression.

    b. What is the answer? If necessary, write the corrected expression to get the correct results when following the order of operations.

32. **Fundraising:** You are on a team that is participating in a charity walk with a goal to raise $12,400. Each team member agrees to raise the same amount of money. If the possible team sizes are 5, 6, 9, or 10 members, which team sizes allow the goal amount to be evenly split between the team members? How much money would each team member raise for each team size that can evenly split the goal amount?

33. **Time:** A company is working on a project that will take 440 hours of work to complete. The manager in charge of the project has the option to have 4, 6, or 8 people work on the project. If the manager wants to evenly divide the work between the team members, which team size will evenly split the work hours? How many hours would each team member spend on the project for each team size that evenly splits the work hours?

34. *Inventory:* Twenty-four pencils are to be distributed evenly between the members of a group. What are the possible group sizes if each person in the group is to receive the same number of pencils?

35. *Baking:* A chocolatier makes 72 specialty truffles. She wants to sell packages that each have the same number of truffles. What are her options for the number of truffles that can be in a package?

36. *Security:* Three security guards meet at the front gate for coffee before they walk around inspecting buildings at a manufacturing plant. The guards take 15, 20, and 30 minutes, respectively, for the inspection trip.

   a. If they start at the same time, in how many minutes will they meet again at the front gate for coffee?

   b. How many trips will each guard have made?

Writing & Thinking

37. Give one example where addition should be completed before multiplication.

38. a. If a number is divisible by both 3 and 5, then it will be divisible by 15. Give two examples.

b. However, a number might be divisible by 3 and not by 5. Give two examples.

c. Also, a number might be divisible by 5 and not 3. Give two examples.

39. Are all odd numbers also prime numbers? Explain your answer.

40. Explain the difference between factors of a number and multiples of that number.

41. Explain, in your own words, why each number in a set divides evenly into the LCM of that set of numbers.

42. Explain why simply multiplying two numbers together will not necessarily find the LCM of those numbers. Give an example of when it would find the LCM and an example when it would not.

1.R.2 Multiplication and Division with Fractions

↻ Making Connections

When solving equations, you will often have to multiply and divide fractions consisting of algebraic expressions and write the answer in lowest terms. Therefore, it is important to review how to multiply fractions and how division of fractions is accomplished through multiplication.

In this section, you will learn skills that you can apply when answering questions like these:

- Solve the following linear equation.

$$\frac{2}{7}(w-1)-\frac{34}{7}=-5$$

- Solve the following rational equation and simplify your answer.

$$\frac{x-4}{6x}=\frac{x+1}{5x}$$

- If Amber were to paint her living room alone, it would take 5 hours. Her sister Heather could do the job in 5 hours. How many hours would it take them working together?

🛠 Building Foundations

To Multiply Fractions

1. _____

2. _____

$$\frac{a}{b} \cdot \frac{c}{d} = \underline{} \quad (b, d \neq \underline{})$$

For example, _____

PROCEDURE

Commutative Property of Multiplication

The order of the fractions being multiplied can be _____

$$\frac{a}{b} \cdot \frac{c}{d} = \underline{} \quad (b, d \neq \underline{})$$

For example, _____

PROPERTIES

20 1.R.2 Multiplication and Division with Fractions

Associative Property of Multiplication

The grouping of the fractions being multiplied can be _____

$$\left(\frac{a}{b} \cdot \frac{c}{d}\right) \cdot \frac{e}{f} = \underline{\hspace{2cm}} \qquad (b, d, f \neq \underline{\hspace{0.5cm}})$$

For example,
_____

PROPERTIES

A fraction is reduced to lowest terms if the numerator and denominator _____

_____

To Reduce a Fraction to Lowest Terms

1. Factor the _____

2. Use the fact that _____

Note: Reduced fractions may be improper fractions.

PROCEDURE

▶ Watch and Work

Watch the video for Example 11 in the software and follow along in the space provided.

Example 11 Multiplying and Reducing Using Prime Factors

Multiply and reduce to lowest terms: $\dfrac{17}{50} \cdot \dfrac{25}{34} \cdot 8$

Solution

✏️ Now You Try It!

Use the space provided to work out the solution to the next example.

Example A Multiplying and Reducing Using Prime Factors

Multiply and reduce to lowest terms:

$16 \cdot \dfrac{12}{100} \cdot \dfrac{5}{36}$

Reciprocals

The reciprocal of $\dfrac{a}{b}$ is _____ . The product of a nonzero number and its reciprocal is

_____

$$\dfrac{a}{b} \cdot \dfrac{b}{a} = 1$$

Note: $0 = \dfrac{0}{1}$, but $\dfrac{1}{0}$ _____

DEFINITION

To Divide Fractions

To divide by any nonzero number, _____

$$\dfrac{a}{b} \div \dfrac{c}{d} = \text{_____}$$

For example, _____

PROCEDURE

▶ Watch and Work

Watch the video for Example 22 in the software and follow along in the space provided.

Example 22 Dividing and Reducing Fractions

Divide and reduce to lowest terms: $\dfrac{16}{27} \div \dfrac{8}{9}$

Solution

✏ Now You Try It!

Use the space provided to work out the solution to the next example.

Example B Dividing and Reducing Fractions

Divide and reduce to lowest terms: $\dfrac{10}{21} \div \dfrac{5}{14}$

🔭 Looking Ahead

The following example requires you to find the least common denominator (LCD) of all the fractions in the equation and then multiply this algebraic expression times every term in the equation and reduce to lowest terms. Note that the resulting equation no longer contains fractions and is a simple linear equation that can easily be solved for the variable.

Example Preview

If Sarah were to paint her living room alone, it would take 5 hours. Her sister Rachel could do the job in 8 hours. How many hours would it take them working together?

Solution

The rate of work for Sarah is $\frac{1}{5}$, while the rate of work for her sister Rachel is $\frac{1}{8}$. If we let x denote the time needed to paint the living room when both sisters are working together, the sum of the two individual rates must equal $\frac{1}{x}$. So, we need to solve the equation $\frac{1}{5} + \frac{1}{8} = \frac{1}{x}$. In this case, the LCD is $40x$.

The equation can be solved as follows.

$$40x \cdot \frac{1}{5} + 40x \cdot \frac{1}{8} = 40x \cdot \frac{1}{x}$$
$$8x + 5x = 40$$
$$13x = 40$$
$$x = \frac{40}{13}$$

It would take them $\frac{40}{13}$, or a little over 3 hours to paint the living room together.

1.R.2 Exercises

Concept Check

True/False. Determine whether each statement is true or false. If a statement is false, explain how it can be changed so the statement will be true. (**Note:** There may be more than one acceptable change.)

1. To find $\frac{1}{2}$ of $\frac{2}{9}$ requires multiplication.

2. $\frac{3}{4} \cdot \frac{9}{10} = \frac{27}{40}$

3. The statement $\frac{1}{3} \cdot \frac{2}{5} = \frac{2}{5} \cdot \frac{1}{3}$ is an example of the associative property of multiplication.

4. The product of a nonzero number and its reciprocal is undefined.

1.R.2 Multiplication and Division with Fractions

5. The reciprocal of 1 is undefined.

6. The result of $\frac{1}{3} \div \frac{1}{6}$ is 2.

7. The reciprocal of 12 is $\frac{12}{1}$.

Practice

Multiply and reduce to lowest terms. (**Hint:** Factor before multiplying.)

8. $\dfrac{0}{3} \cdot \dfrac{5}{7}$

9. $\dfrac{1}{3} \cdot \dfrac{3}{4}$

10. $\left(-\dfrac{1}{5}\right)\left(-\dfrac{4}{7}\right)$

11. $\dfrac{5}{16} \cdot \dfrac{16}{15}$

12. $\dfrac{9}{10} \cdot \dfrac{35}{40} \cdot \dfrac{25}{15}$

Divide and reduce to lowest terms.

13. $\dfrac{2}{3} \div \dfrac{3}{4}$

14. $0 \div \dfrac{5}{6}$

15. $\dfrac{5}{6} \div 0$

16. $\dfrac{14}{15} \div \dfrac{21}{25}$

Applications

Solve.

17. **Recipes:** A recipe calls for $\frac{3}{4}$ cups of flour. How much flour should be used if only half of the recipe is to be made?

18. **Demographics:** A study showed that $\frac{3}{5}$ of the students in an elementary school were left-handed. If the school had an enrollment of 600 students, how many were left-handed?

19. **Geology:** The floor of the Atlantic Ocean is spreading apart at an average rate of $\frac{3}{50}$ of a meter per year. How long will it take for the ocean floor to spread 12 meters?

20. **Airplane Capacity:** An airplane is carrying 180 passengers. This is $\frac{9}{10}$ of the capacity of the airplane.
    a. Is the capacity of the airplane more or less than 180?

    b. If you were to multiply 180 times $\frac{9}{10}$, would the product be more or less than 180?

    c. What is the capacity of the airplane?

Writing & Thinking

21. If two fractions are between 0 and 1, can their product be more than 1? Explain.

22. Explain the process of multiplying two fractions. Give an example of a product that cannot be reduced.

23. Explain why the number 0 has no reciprocal.

24. Is division a commutative operation? Explain briefly and give three examples using fractions to help justify your answer.

1.R.3 Addition and Subtraction with Fractions

♻ Making Connections

When solving rational equations, you will often have to add and subtract fractions containing algebraic expressions and write the answer in lowest terms. Therefore, a review of how to add and subtract numerical fractions, particularly when the fractions have unlike denominators would be beneficial.

In this section, you will learn skills that you can apply when answering questions like these:

- Solve the following rational equation and simplify your answer.

$$\frac{2x}{x+2} + \frac{3}{x+1} = 2$$

- Solve the following rational equation and simplify your answer.

$$\frac{x}{x-3} - \frac{8}{x+5} = \frac{x^2}{x^2+2x-15}$$

- Miguel can drive 4 times as fast as Oscar can ride his bicycle. If it takes Oscar 3 hours longer than Miguel to travel 48 miles, how fast (in mph) can Oscar ride his bicycle?

🛠 Building Foundations

To Add Fractions with the Same Denominator

1. _____

2. _____ $\dfrac{a}{b} + \dfrac{c}{b} = $ _____

3. _____

For example, _____

PROCEDURE

To Add Fractions with Different Denominators

1. Find the _____

2. Change each fraction into _____

3. _____

4. _____

PROCEDURE

1.R.3 Addition and Subtraction with Fractions

Commutative Property of Addition
The order of the fractions being added can be _____

$$\frac{a}{b} + \frac{c}{d} = \underline{\hspace{2cm}}$$

For example, _____

PROPERTIES

Associative Property of Addition
The grouping of the fractions being added can be _____

$$\frac{a}{b} + \left（\frac{c}{d} + \frac{e}{f}\right) = \underline{\hspace{2cm}}$$

For example, _____

PROPERTIES

To Subtract Fractions with the Same Denominator
1. _____
2. _____
3. _____

$$\frac{a}{b} - \frac{c}{b} = \underline{\hspace{2cm}}$$

For example, _____

PROCEDURE

To Subtract Fractions with Different Denominators
1. Find the _____
2. Change each fraction into _____
3. _____
4. _____

PROCEDURE

▶ Watch and Work

Watch the video for Example 11 in the software and follow along in the space provided.

Example 11 Subtracting Fractions with Different Denominators

Subtract: $1 - \dfrac{5}{8}$

Solution

✏ Now You Try It!

Use the space provided to work out the solution to the next example.

Example A Subtracting Fractions with Different Denominators

Subtract: $3 - \dfrac{5}{12}$

1.R.3 Addition and Subtraction with Fractions

🔭 Looking Ahead

The following example shows one method for solving rational equations. This method requires the subtraction of rational expressions, which follows the same procedure as the subtraction of fractions.

Example Preview

Solve the following rational equation and simplify your answer.

$$\frac{x}{x+2} - \frac{1}{x-4} = 1$$

Solution

To solve this equation, assume no denominator is 0. This means $x \neq -2, 4$. Begin by subtracting the rational expressions on the left side of the equation. Note that the LCD is $(x+2)(x-4)$. Change each expression into an equivalent rational expression with that denominator and subtract.

$$\frac{x}{x+2} \cdot \frac{x-4}{x-4} - \frac{1}{x-4} \cdot \frac{x+2}{x+2} = 1$$

$$\frac{x^2 - 4x - (x+2)}{(x+2)(x-4)} = 1$$

$$\frac{x^2 - 4x - x - 2}{x^2 - 2x - 8} = 1$$

$$\frac{x^2 - 5x - 2}{x^2 - 2x - 8} = 1$$

Next, multiply both sides by the denominator and solve the resulting equation.

$$x^2 - 5x - 2 = x^2 - 2x - 8$$
$$-5x - 2 = -2x - 8$$
$$-3x = -6$$
$$x = 2$$

Since 2 is not equal to -2 or 4, the solution is $\{2\}$.

1.R.3 Exercises

Concept Check

True/False. Determine whether each statement is true or false. If a statement is false, explain how it can be changed so the statement will be true. (**Note:** There may be more than one acceptable change.)

1. The final step in adding fractions is to reduce, if possible.

2. The process for finding the LCD is the same as the process for finding the LCM.

3. LCD represents the Least Common Digit.

4. When subtracting fractions, simply subtract the numerators and the denominators.

5. Subtraction of fractions requires that the fractions have the same denominators.

Practice

Add and reduce to lowest terms.

6. $\dfrac{3}{25} + \dfrac{12}{25}$

7. $\dfrac{2}{7} + \dfrac{4}{21} + \dfrac{1}{3}$

Subtract and reduce to lowest terms.

8. $-\dfrac{7}{15} + \dfrac{3}{5}$

11. $\dfrac{9}{14} - \dfrac{2}{21}$

9. $\dfrac{1}{4} + \left(-\dfrac{1}{20}\right) + \dfrac{8}{15}$

12. $2 - \dfrac{9}{16}$

10. $\dfrac{7}{8} - \dfrac{5}{8}$

13. $-\dfrac{5}{12} - \left(-\dfrac{1}{6}\right)$

Applications

Solve.

14. **Cooking:** A recipe calls for the following spices: $\frac{1}{2}$ teaspoon of turmeric, $\frac{1}{4}$ teaspoon of ginger, and $\frac{1}{8}$ teaspoon of cumin. What is the total quantity of these three spices?

15. **Postage:** Three pieces of mail weigh $\frac{1}{2}$ ounce, $\frac{1}{5}$ ounce, and $\frac{3}{10}$ ounce. What is the total weight of the letters?

Writing & Thinking

16. Give an example of a situation where you might add or subtract fractions (other than in class).

17. Explain how finding the LCM relates to LCDs.

1.R.4 Proportions

♻ Making Connections

You will often have to solve equations involving rational expressions. Therefore, it is important to review how to solve a proportion by multiplying both sides of the equation by the least common denominator.

In this section, you will learn skills that you can apply when answering questions like these:

- Solve the following rational equation.

$$\frac{3}{x-2}+\frac{2}{x+1}=1$$

- In order to flush deposits from a radiator, a drain that can empty the entire radiator in 45 minutes is left open at the same time it is being filled at a rate that would fill it in 30 minutes. How long does it take for the radiator to fill?

🛠 Building Foundations

Proportions

A **proportion** is a statement that _____

In symbols,
_____

A proportion is true if _____

DEFINITION

To Solve a Proportion

1. Find the cross products (or cross multiply) and then _____

2. Divide both sides of the equation by _____

3. _____

PROCEDURE

1.R.4 Proportions

▶ Watch and Work

Watch the video for Example 3 in the software and follow along in the space provided.

Example 3 Solving Proportions

Find the value of x if $\dfrac{4}{8} = \dfrac{5}{x}$.

Solution

✏ Now You Try It!

Use the space provided to work out the solution to the next example.

Example A Solving Proportions

Find the value of x if $\dfrac{12}{x} = \dfrac{9}{15}$.

To Solve an Application Using a Proportion

1. Identify the unknown quantity and _____

2. Set up a proportion in which the _____. (Make sure that the _____)

3. Solve the _____

PROCEDURE

👀 Looking Ahead

The following example requires you to find the least common denominator (LCD) of the rational expressions in the proportional equation and then multiply each side of the equation by this denominator and simplify. Note that the resulting equation no longer contains fractions and becomes a quadratic equation that can be solved by factoring.

Example Preview

$$\frac{z^3 - 5z^2}{z^2 + 4z - 45} = \frac{-13z - 36}{z + 9}$$

Solution

The first thing to do factor all numerators and denominators in this equation.

$$\frac{z^2(z-5)}{(z+9)(z-5)} = \frac{-13z - 36}{z + 9}$$

At this point, it must be noted that both 5 and −9 must be excluded as possible solutions for this rational equation.

Before determining the LCD, there is a common factor of $(z-5)$ in the numerator and denominator of the fraction on the left side of the equation that needs to be canceled.

$$\frac{z^2 \cancel{(z-5)}}{(z+9)\cancel{(z-5)}} = \frac{-13z - 36}{z + 9}$$

$$\frac{z^2}{z+9} = \frac{-13z - 36}{z + 9}$$

It should now be clear that the LCD is $(z+9)$.

Multiplying both sides of the equation by the LCD results in the following.

$$(z+9)\frac{z^2}{z+9} = (z+9)\frac{-13z-36}{z+9}$$

$$z^2 = -13z - 36$$

$$z^2 + 13z + 36 = 0$$

The resulting quadratic equation can be solved by factoring.

$$z^2 + 13z + 36 = 0$$
$$(z+9)(z+4) = 0$$
$$z = -4, \cancel{-9}$$

Since it was determined earlier that −9 must be excluded as a solution to this rational equation, the solution is −4.

1.R.4 Exercises

Concept Check

True/False. Determine whether each statement is true or false. If a statement is false, explain how it can be changed so the statement will be true. (**Note:** There may be more than one acceptable change.)

1. A proportion is a statement that two ratios are being multiplied.

2. Cross canceling is used to determine if a proportion is true.

3. In order to solve the proportion $\frac{16}{36.8} = \frac{x}{27.6}$ we construct the equation $36.8x = 441.6$.

4. When using proportions to solve a word problem, there is only one correct way to set up the proportion.

5. The proportions $\frac{36 \text{ tickets}}{\$540} = \frac{x \text{ tickets}}{\$75}$ and $\frac{x \text{ tickets}}{36 \text{ tickets}} = \frac{\$75}{\$540}$ will yield the same answer.

Practice

Determine whether each proportion is true or false.

6. $\dfrac{3}{6} = \dfrac{4}{8}$

7. $\dfrac{1}{3} = \dfrac{33}{100}$

Solve each proportion.

8. $\dfrac{5}{4} = \dfrac{x}{8}$

9. $\dfrac{3.5}{2.6} = \dfrac{10.5}{B}$

Applications

Solve.

10. **Concrete:** The quality of concrete is based on the ratio of bags of cement to cubic yards of gravel. One batch of concrete consists of 27 bags of cement mixed into 9 cubic yards of gravel, while a second has 15 bags of cement mixed with 5 cubic yards of gravel. Determine whether the ratio of cement to gravel is the same for both batches.

11. **Grading:** An English teacher must read and grade 27 essays. If the teacher takes 20 minutes to read and grade 3 essays, how much time will he need to grade all 27 essays?

Writing & Thinking

12. In your own words, clarify how you can know that a proportion is set up correctly or not.

1.R.5 Decimals, Fractions, and Percents

↻ Making Connections

When simplifying algebraic expressions or solving algebraic equations, you will often have to work with decimals and percents. Therefore, it is important to understand the relationship between decimals and fractions and percents.

In this section, you will learn skills that you can apply when answering questions like these:

- Natalie's check at Aubrey's Kitchen is $19.64. In order to leave an 18% tip, how much should she pay?
- A tool shed costs $7070 before sales tax. Paige pays $7352.80 for the tool shed, including tax. What is the tax rate where Paige lives?

🛠 Building Foundations

The word percent comes from the Latin *per centum*, meaning _____. So **percent means** _____, or **the ratio of a number to** _____.

The symbol % is called the _____. This sign has the same meaning as the fraction $\frac{1}{100}$.

To Change a Decimal Number to a Percent

1. Move the _____
2. Write the _____

PROCEDURE

To Change a Percent to a Decimal Number

1. Move the _____
2. Delete the _____

PROCEDURE

▶ Watch and Work

Watch the video for Example 3 in the software and follow along in the space provided.

Example 3 Changing Percents to Decimal Numbers

Change each percent to a decimal number.

a. 76%

b. 18.5%

c. 50%

d. 100%

e. 0.25%

✏ Now You Try It!

Use the space provided to work out the solution to the next example.

Example A Changing Percents to Decimal Numbers

Change each percent to a decimal number.

a. 40%   d. 29.37%

b. 211%   e. 102%

c. 0.6%

1.R.5 Decimals, Fractions, and Percents

To Change a Fraction to a Percent
1. Change the _____. (Divide _____.)
2. Change the _____.

PROCEDURE

▶ Watch and Work

Watch the video for Example 6 in the software and follow along in the space provided.

Example 6 Changing Mixed Numbers to Percents

Change $2\frac{1}{4}$ to a percent.

Solution

✏ Now You Try It!

Use the space provided to work out the solution to the next example.

Example B Changing Mixed Numbers to Percents

Change $1\frac{1}{2}$ to a percent.

> **To Change a Percent to a Fraction or a Mixed Number**
> 1. Write the percent as a fraction with _____
> 2. Reduce the _____
>
> PROCEDURE

👀 Looking Ahead

Your review of percents will be helpful in the following example, which involves calculating the tip at a restaurant.

Example Preview

Marvin decides to leave a 15% tip after eating dinner at Fresh Catchery. If the bill is $27.32, how much should he pay?

Solution

First, find the tip amount by finding 15% of $27.32. Converting 15% to a decimal gives 0.15. The tip amount is $(0.15)(\$27.32) = \$4.098$, which rounds to $4.10.

The total cost is thus $27.32 + $4.10 = $31.42. Marvin should pay $31.42.

1.R.5 Exercises

Concept Check

True/False. Determine whether each statement is true or false. If a statement is false, explain how it can be changed so the statement will be true. (**Note:** There may be more than one acceptable change.)

1. If a decimal number is less than 1, then the equivalent percent will be less than 100%.

2. It is not possible to have a percent greater than 100%.

3. A decimal number that is between 0.01 and 0.10 is between 10% and 100%.

4. To change from a percent to a decimal, simply omit the percent sign.

5. Fractions that have denominators other than 100 cannot be changed to a percent.

6. The fraction $\frac{1}{5}$ is equivalent to $\frac{1}{5}$%.

Practice

Change each fraction to a percent.

7. $\frac{20}{100}$

8. $\frac{125}{100}$

Change each decimal number to a percent.

9. 0.02

10. 2.3

Change each percent to a decimal number.

11. 7%

12. 179%

Change each fraction or mixed number to a percent. If necessary, round to the nearest tenth of a percent.

13. $\frac{3}{4}$

14. $5\frac{3}{10}$

Change each percent to a fraction or mixed number and reduce, if possible.

15. 120%

16. 12.5%

Applications

Solve.

17. *Interest:* A savings account is offering an interest rate of 0.04 for the first year after opening the account. Change 0.04 to a percent.

18. *Sales Tax:* Suppose that sales tax is figured at 7.25%. Change 7.25% to a decimal.

19. *Exam Grades:* Out of a possible total of 240 points on an exam, David received 204 points. What percent of the exam did David get correct?

20. *College Degrees:* To receive a Bachelor of Science (BS) degree at Bluefield State College, the student must complete a total of 128 credit hours, of which 41 of these credits must be general education Core Skills courses. What percent of the total curriculum is dedicated to general education courses? [1]

Writing & Thinking

21. Describe the relationship between percent and the number 100.

22. Describe a situation where more than 100% is possible. Describe a situation where it is impossible to have more than 100%.

---

1 Source: 2010–2011 Bluefield, WV State College Catalogue, p.76

1.R.5 Decimals, Fractions, and Percents

23. Justify why mixed numbers are a larger percentage than proper fractions alone. (Consider the value of 100%.)

24. Describe the process to change a percent to a fraction or mixed number.

CHAPTER 2.R

Review Concepts

for *Equations and Inequalities in One Variable*

2.R.1 The Real Number Line and Absolute Value

2.R.2 Addition with Real Numbers

2.R.3 Subtraction with Real Numbers

2.R.4 Multiplication and Division with Real Numbers

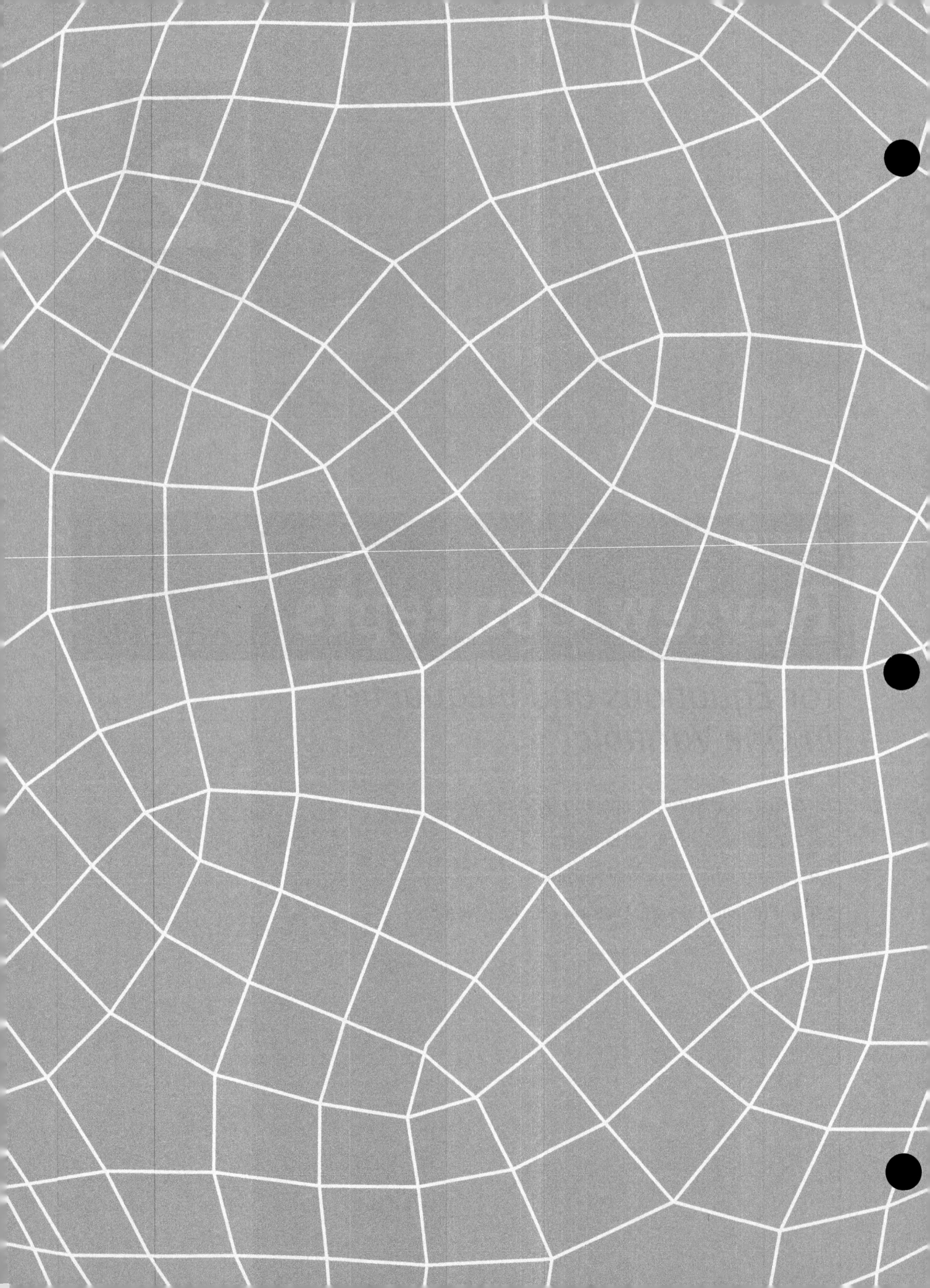

2.R.1 The Real Number Line and Absolute Value

↻ Making Connections

Most quantities we encounter in our daily lives, such as a bank account balance or the temperature outside, can be measured and represented using real numbers. The real number line allows us to visualize and compare such values.

In this section, you will learn skills that you can apply when answering questions like these:

- Select all the symbols from the set $\{>, \geq, \leq, <\}$ that can be placed in the blank to make the following a true statement.

$$\frac{1}{4} \underline{\qquad} \frac{1}{5}$$

- Evaluate the following expression for the given values of the variables.

$$|x - 7y| + (8z - 4); \text{ for } x = -2, y = -1, \text{ and } z = 2$$

- Solve the following inequality and express your answer in interval notation. Then graph the solution set.

$$10z - 8 < 9 + 12z$$

🛠 Building Foundations

Integers
The set of numbers consisting of the _____
_____
DEFINITION

Variables
A **variable** is a symbol (generally a _____) that is used to _____
_____
DEFINITION

2.R.1 The Real Number Line and Absolute Value

> **Rational Numbers**
>
> A **rational number** is a number that can be written in _____
>
> _____
>
> OR
>
> A **rational number** is a number that can be written in _____
>
> _____
>
> DEFINITION

Numbers that cannot be written as fractions with integer numerators and denominators are called _____

▶ Watch and Work

Watch the video for Example 4 in the software and follow along in the space provided.

Example 4 Graphing Sets of Numbers

Graph the set of **real numbers** $\left\{-\frac{3}{4}, 0, 1, 1.5, 3\right\}$.

Solution

✏ Now You Try It!

Use the space provided to work out the solution to the next example.

Example A Graphing Sets of Numbers

Graph the set of real numbers
$$\left\{-2.5, -1, 0, \frac{5}{4}, 4\right\}.$$

Symbols of Equality and Inequality

Reading from left to right:

= _____   ≠ _____

< _____   > _____

≤ _____   ≥ _____

Absolute Value

The **absolute value** of a real number is _____ Note that the absolute value of a _____

PROPERTIES

🔭 Looking Ahead

Your knowledge of the definition of absolute value will help you solve equations involving the concept. Absolute value equations can usually be rewritten as two distinct equations, as you will see in this next example.

Example Preview

Solve the following absolute value equation.

$$|6y+5|=16$$

Solution

$$6y+5-16 \qquad 6y+5=-16$$
$$6y=11 \qquad\quad 6y=-21$$
$$\text{or}$$
$$y=\frac{11}{6} \qquad\quad y=-\frac{7}{2}$$

Since both of these solutions solve the original equation, this absolute value equation has two solutions.

2.R.1 Exercises

Concept Check

True/False. Determine whether each statement is true or false. If a statement is false, explain how it can be changed so the statement will be true. (**Note:** There may be more than one acceptable change.)

1. On a number line, smaller numbers are always to the left of larger numbers.

2. The absolute value of a negative number is a positive number.

3. All whole numbers are also integers.

4. Zero is a positive number.

Practice

Graph each set of real numbers on a real number line.

5. $\{-3, -2, 0, 1\}$

6. $\left\{-2, -1, -\dfrac{1}{3}, 2\right\}$

List the numbers in the set $A = \left\{-7, -\sqrt{6}, -2, -\dfrac{5}{3}, -1.4, 0, \dfrac{3}{5}, \sqrt{5}, \sqrt{11}, 4, 5.9, 8\right\}$ that are described in each exercise.

7. Whole numbers

8. Rational numbers

Determine whether each statement is true or false. If a statement is false, rewrite it in a form that is a true statement. (There may be more than one way to correct a statement.)

9. $0 = -0$

10. $|-8| \geq 4$

Applications

Solve. Represent each quantity with a signed integer.

11. *Oceans:* The Alvin is a manned deep-ocean research submersible that has explored the wreck of the Titanic. The operating depth of the Alvin is 4500 meters below sea level.

12. *Oceans:* The Mariana trench is the deepest known location on the Earth's ocean floor. The deepest known part of the Mariana Trench is approximately 11 kilometers below sea level.

Writing & Thinking

13. Explain, in your own words, how an expression such as $-y$ might represent a positive number.

14. Compare and contrast absolute value with opposites.

2.R.2 Addition with Real Numbers

↻ Making Connections

Addition is one of the four main operations involving the real numbers; the others being subtraction, multiplication, and division. We use addition in our everyday lives, like when we shop, cook, or budget our monthly expenses. Addition is also very important when performing computations and solving algebra problems.

In this section, you will learn skills that you can apply when answering questions like this:

- Evaluate the following expression.

$$\frac{(2+4\cdot 6-5)}{-8(4-6\div(-1+8))}$$

- Solve the following linear equations.

$$4x-4=-2$$

$$\frac{8y-5}{5}+\frac{17}{10}=\frac{16y+7}{10}$$

🛠 Building Foundations

Rules for Addition with Real Numbers

1. To add two real numbers with **like signs**,

   a. _____

   b. use the _____

2. To add two real numbers with **unlike signs**,

   a. _____

   b. use the _____

PROCEDURE

▶ Watch and Work

Watch the video for Example 3 in the software and follow along in the space provided.

Example 3 Adding Three or More Real Numbers

Add.

a. $-3 + 2 + (-5)$
b. $6.0 + (-4.3) + (-1.5)$

Solution

✏ Now You Try It!

Use the space provided to work out the solution to the next example.

Example A Adding Three or More Real Numbers

Add.

a. $-7 + 5 + (-3)$
b. $-3.2 + (-6.1) + 5.7$

🔭 Looking Ahead

Being able to accurately add both negative and positive real numbers is a fundamental skill needed to solve linear equations in one variable. We want to be able to find the value of the variable that makes the proposed equation a true statement. You may need to add both negative and positive numbers to both sides of the equation in order to balance it.

Example Preview

Solve the following linear equation.

$$-13 = -3u - 16$$

Solution

$$-13 = -3u - 16$$
$$-13 + 16 = -3u - 16 + 16$$
$$3 = -3u$$
$$\frac{3}{-3} = \frac{-3u}{-3}$$
$$-1 = u$$

2.R.2 Exercises

Concept Check

True/False. Determine whether each statement is true or false. If a statement is false, explain how it can be changed so the statement will be true. (**Note:** There may be more than one acceptable change.)

1. The sum of a positive number and a negative number is always positive.

2. When adding two numbers with unlike signs, the result uses the sign of the number with the larger absolute value.

3. The sum of two positive numbers can equal zero.

Practice

Add. Reduce any fractions to lowest terms.

4. $8+(-3)$

5. $2+(-8)$

6. $-\dfrac{1}{6}+\dfrac{7}{15}$

7. $3.2+(-1.2)+(-2.5)$

Add. Be sure to find the absolute values first.

8. $13+|-5|$

Applications

Solve.

9. **Profit:** For 2017, a business reports a profit of $45,000 during the first quarter, a loss of $8000 during the second quarter, a loss of $2000 during the third quarter, and a profit of $15,000 during the fourth quarter.

   a. Write an addition expression to represent the total profit made by the company in 2017. Do not simplify.

   b. Simplify the expression from Part **a.**

10. **Oceans:** A submarine dives to a depth of 250 feet below the surface. It rises 75 feet before diving an additional 100 feet. What is the final depth of the submarine?

Writing & Thinking

11. Describe, in your own words, how the sum of the absolute values of two numbers might be 0. (Is this even possible?)

12. Describe in your own words the conditions under which the sum of two integers will be 0.

2.R.2 Addition with Real Numbers

2.R.3 Subtraction with Real Numbers

↻ Making Connections

We often need to know how much a quantity has changed over time, like when the temperature outside increases or when you make a payment on a student loan. The operation needed is the subtraction of real numbers, which is also needed when simplifying algebraic expressions.

In this section, you will learn skills that you can apply when answering questions like these:

- Simplify the following expression.

$$(15 - 15x) - (12)$$

- Subtract the the two rational expressions and reduce your answer to lowest terms.

$$\frac{x+2}{x-3} - \frac{x-3}{x-2}$$

🛠 Building Foundations

Additive Inverse

The opposite of a _____. The sum of a number and its

additive inverse _____. Symbolically, for any real number a,

_____

DEFINITION

Subtraction

For any real numbers a and b,

_____

In words, to subtract b from a, _____

DEFINITION

To find the **change in value** between two numbers, _____

_____. Symbolically, change in value = _____.

▶ Watch and Work

Watch the video for Example 4 in the software and follow along in the space provided.

Example 4 Application: Calculating Change in Value

A jet pilot flew her plane from an altitude of 30,000 ft to an altitude of 12,000 ft. What was the change in altitude?

Solution

✏ Now You Try It!

Use the space provided to work out the solution to the next example.

Example A Calculating Change in Value

A drone plane flew from an altitude of 25,000 ft to an altitude of 14,000 ft. What was the change in altitude?

Looking Ahead

Reviewing the main ideas related to subtraction of real numbers will allow you to understand and simplify problems involving the addition and subtraction of complex numbers.

Example Preview

Simplify the following expression.

$$(12-18i)-(5-12i)$$

Solution

$$(12-18i)-(5-12i) = (12-5)+(-18-(-12))i$$
$$= 7-6i$$

2.R.3 Exercises

True/False. Determine whether each statement is true or false. If a statement is false, explain how it can be changed so the statement will be true. (**Note:** There may be more than one acceptable change.)

1. The sum of a number and its additive inverse is the number itself.

2. The additive inverse of negative seven is seven.

3. We can think of addition of numbers as accumulating numbers.

4. The expression "15 − 7" can be thought of as "fifteen plus negative seven."

Practice

Find the additive inverse (opposite) of each real number.

5. 11

6. −3.4

2.R.3 Subtraction with Real Numbers

Subtract. Reduce fractions to lowest terms.

7. $-8-(-11)$

8. $\dfrac{7}{15} - \dfrac{2}{15}$

Perform the indicated operation to find the net change in value.

9. $-6+(-4)-5$

10. $-11.3 + 5.3 - 7.9$

Applications

Solve.

11. **Temperature:** At 2 p.m. the temperature was 76 °F. At 8 p.m. the temperature was 58 °F. What was the change in temperature?

12. **Real Estate:** A couple sold their house for $135,000. They paid the realtor $8100, and other expenses of the sale came to $800. If they owed the bank $87,000 for the mortgage, what were their net proceeds from the sale?

Writing & Thinking

13. Explain, in your own words, how to find the difference between a positive and a negative number.

14. What is the additive inverse of 0? Why?

2.R.4 Multiplication and Division with Real Numbers

↻ Making Connections

Multiplication and division are two very important operations in the set of real numbers. They allow us to repeatedly add and subtract in a very efficient way. Mastery of such concepts is very important when performing computations with exponents as well as tackling problems involving rational expressions and polynomials.

In this section, you will learn skills that you can apply when answering questions like these:

- Perform the indicated operation.

$$(5.96)(-3.7)$$

- Multiply the following polynomials

$$(4x-5y)(x+3y)$$

✂ Building Foundations

> ### Rules for Multiplication with Real Numbers
> If a and b are positive real numbers, then
> 1. The product of two positive numbers _____
> 2. The product of two negative numbers _____
> 3. The product of a positive number and a negative number _____
> 4. The product of 0 and any number _____
>
> **PROCEDURE**

> ### Division with Real Numbers
> For real numbers a, b, and x (where $b \neq 0$),
>
> $$\frac{a}{b} = x \text{ means } _____$$
>
> For real numbers a and b (where $b \neq 0$),
>
> $$\frac{a}{0} \text{ is } _____$$
>
> **DEFINITION**

2.R.4 Multiplication and Division with Real Numbers

> **Rules for Division with Real Numbers**
>
> If a and b are positive real numbers (where $b \neq 0$),
>
> 1. The quotient of two positive numbers _____
>
> 2. The quotient of two negative numbers _____
>
> 3. The quotient of a positive number and a negative number _____
>
> **PROCEDURE**

> **Average**
>
> The **average** (or **mean**) of a set of numbers is the value found by _____
> _____
>
> **DEFINITION**

▶ Watch and Work

Watch the video for Example 6 in the software and follow along in the space provided.

Example 6 Application: Calculating an Average

At noon on five consecutive days in Aspen, Colorado the temperatures were −5°, 7°, 6°, −7°, and 14° (in degrees Fahrenheit). (Negative numbers represent temperatures below zero). Find the average of these noonday temperatures.

Solution

✏️ Now You Try It!

Use the space provided to work out the solution to the next example.

Example A Calculating an Average

At noon on five consecutive days in Mears, Michigan the temperatures were −3°, 5°, 8°, −4°, and 14° (in degrees Fahrenheit.) (Negative numbers represent temperatures below zero.) Find the average of these noonday temperatures.

🔭 Looking Ahead

Now that you have reviewed and mastered the multiplication and division of real numbers, you will be able to apply these ideas to concrete problems involving investments, as in this example.

Example Preview

Mercedes has $2800 that she wants to invest in a savings account for 3.7 years, at which time she plans to close out the account and use the money as a down payment on a house. She finds one local bank offering an annual interest rate of 2.3% compounded daily (Bank 1), and another bank offering an annual interest rate of 2.4% compounded semiannually (Bank 2). Which bank should she choose?

Solution

$$A(t) = P\left(1 + \frac{r}{n}\right)^{nt}$$

$$A_1(3.7) = 2800\left(1 + \frac{0.023}{365}\right)^{365(3.7)}$$
$$\approx 2800(1.000063014)^{1350.5}$$
$$\approx 3048.70$$

or

$$A_2(3.7) = 2800\left(1 + \frac{0.024}{2}\right)^{2(3.7)}$$
$$= 2800(1.012)^{7.4}$$
$$\approx 3058.40$$

From the results, we see that Bank 2 offers the better interest rate

2.R.4 Exercises

True/False. Determine whether each statement is true or false. If a statement is false, explain how it can be changed so the statement will be true. (**Note:** There may be more than one acceptable change.)

1. If a negative number is divided by a positive number, the result will be a negative number.

2. The product of zero and a number is zero.

3. If two numbers have the same sign, both the product and the quotient of the two numbers will be negative.

4. The mean of a set of numbers is always positive.

Practice

Multiply. Reduce fractions to lowest terms.

5. $12 \cdot 4$

6. $(-7)(-16)(0)$

Divide. Reduce fractions to lowest terms. Round answers with decimals to the nearest tenth.

7. $\dfrac{-20}{-10}$

8. $\dfrac{-5.6}{7}$

Applications

Solve.

9. *Mean:* Find the mean of the following set of integers: −10, 15, 16, −17, −34, and −42.

10. *Animals:* According to the US Fish and Wildlife Service, migratory birds are imported at a value of about $19 each. Suppose that about 800,000 live birds are imported each year. What is the total value of these imported birds?

Writing & Thinking

11. If you multiply an odd number of negative numbers together, do you think that the product will be positive or negative? Explain your reasoning.

12. Explain the conditions under which the quotient of two numbers is 0.

2.R.4 Multiplication and Division with Real Numbers

CHAPTER 3.R

Review Concepts

for *Equations and Inequalities in Two Variables*

3.R.1 Formulas in Geometry

3.R.2 Square Roots and the Pythagorean Theorem

3.R.3 Evaluating Radicals

3.R.4 Simplifying Radicals

3.R.5 Introduction to the Cartesian Coordinate System

3.R.6 Solving Linear Equations: $ax + b = c$

3.R.7 Solving Linear Equations: $ax + b = cx + d$

3.R.8 Solving Linear Inequalities in One Variable

3.R.9 Solving Radical Equations

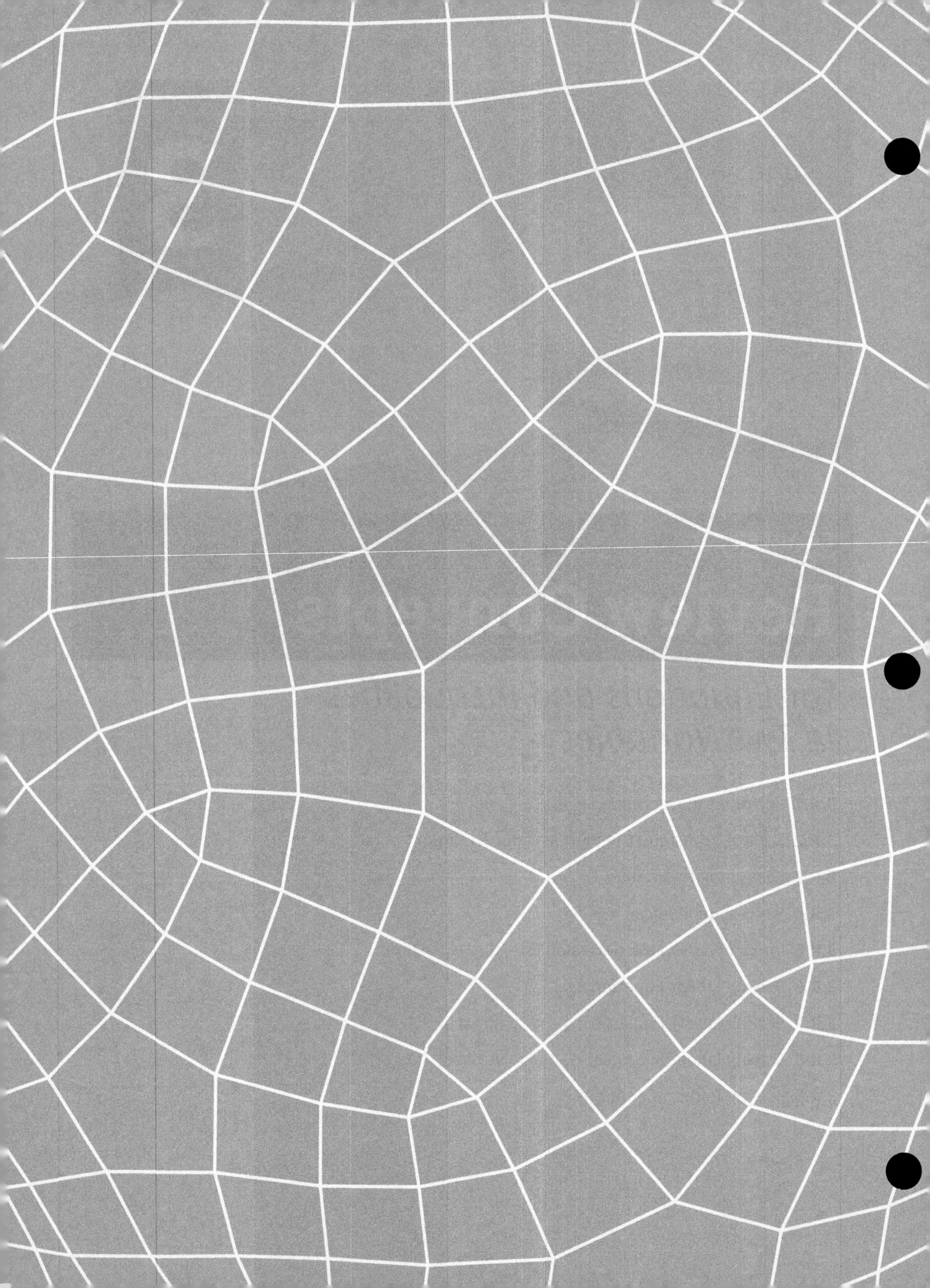

3.R.1 Formulas in Geometry

⟳ Making Connections

Many of the real world problems you encounter will contain some type of geometric figure in their conception and/or solution. Therefore, to better prepare yourself to handle problems involving geometric shapes, it is recommended that you review common vocabulary and formulas from geometry.

In this section, you will learn skills that you can apply when answering questions like these:

- Prove that the triangle with vertices at the points (1,1), (−2,−5), and (3,0) is a right triangle. Then determine the area of the triangle.

- Determine if the quadrilateral graphed below is a parallelogram.

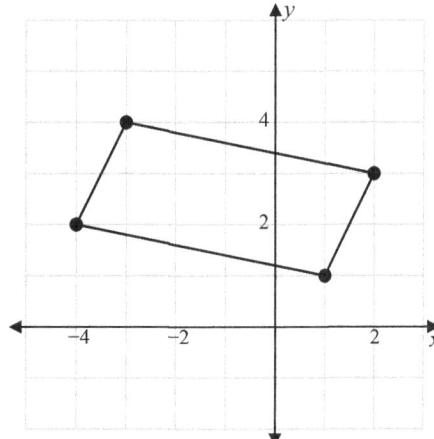

🛠 Building Foundations

Perimeter

> **Polygon**
>
> A **polygon** is a closed plane figure, with _____
>
> Each point where _____
>
> **Note:** A **closed figure** begins and ends at the same point.
>
> **DEFINITION**

A **triangle** is _____

A **parallelogram** is _____

_____

A **rectangle** is _____

3.R.1 Formulas in Geometry

A **square** is _____

A **trapezoid** is _____

> **Perimeter**
>
> The **perimeter** P of a polygon is _____
>
> DEFINITION

> **Perimeter Formulas for Five Polygons**
>
>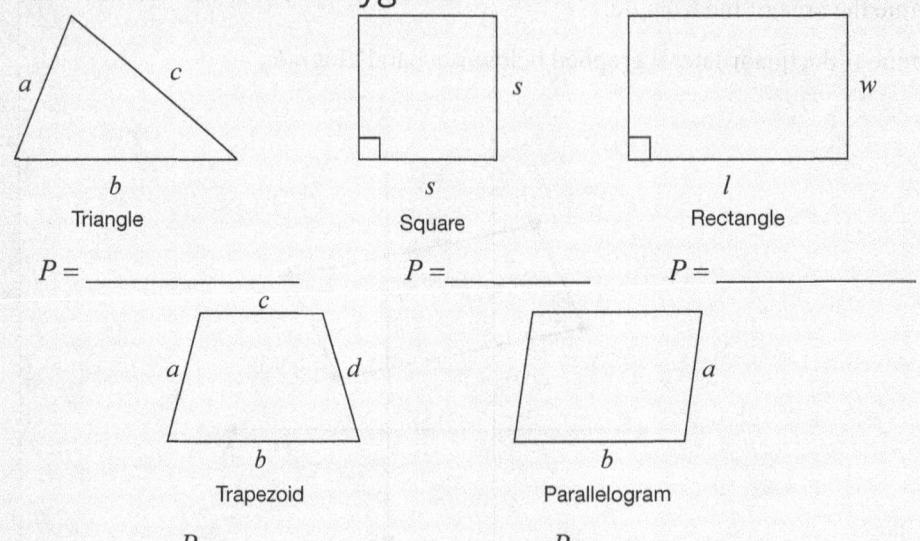
>
> Triangle   Square   Rectangle
>
> $P = $ _____ $P = $ _____ $P = $ _____
>
> Trapezoid   Parallelogram
>
> $P = $ _____ $P = $ _____
>
> Note that each formula represents the sum of the lengths of the sides.
>
> FORMULA

▶ Watch and Work

Watch the video for Example 2 in the software and follow along in the space provided.

Example 2 Calculating the Perimeter of a Triangle

Calculate the perimeter of a triangle with sides of length 40 mm, 70 mm, and 80 mm.

Solution

✏️ Now You Try It!

Use the space provided to work out the solution to the next example.

Example A Application: Finding the Mean

Calculate the perimeter of a triangle with sides of length 15 ft, 40 ft, and 30 ft. It may be helpful to begin by drawing the figure.

Area

Area is a measure of _____.

Area is measured in _____.

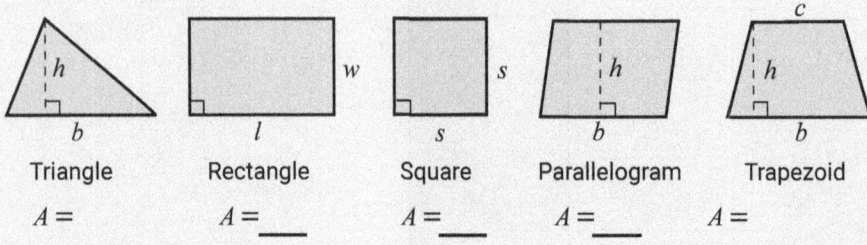

Area Formulas for Five Polygons

Triangle A = _____

Rectangle A = _____

Square A = _____

Parallelogram A = _____

Trapezoid A = _____

Note: The letter h is used to represent the **height** of the figure. The height is also called the **altitude** and is perpendicular to the base.

FORMULA

▶ Watch and Work

Watch the video for Example 8 in the software and follow along in the space provided.

Example 8 Calculating the Area of a Trapezoid Using a Formula

Calculate the area of a trapezoid with altitude 6 in. and parallel sides of length 12 in. and 24 in.

Solution

Now You Try It!

Use the space provided to work out the solution to the next example.

Example B **Calculating the Area of a Trapezoid Using a Formula**

Calculate the area of a trapezoid with altitude 3 cm and parallel sides of length 9 cm and 15 cm. It may be helpful to begin by drawing the figure.

Circles

> **Circles**
>
> **Circle:** The set of all points in a plane that are _____
> _____
>
> **Radius:** The distance from the _____
>
> (_____ is used to represent the radius of a circle.)
>
> **Diameter:** The distance from _____
> _____
>
> (_____ is used to represent the diameter of a circle and _____.)
>
> **Circumference:** _____
>
> DEFINITION

3.R.1 Formulas in Geometry

Formulas for Circles

Circumference: $C = $ _____ and $C = $ _____

Area: $A = $ _____

FORMULA

▶ Watch and Work

Watch the video for Example 13 in the software and follow along in the space provided.

Example 13 Calculating the Circumference and Area of a Circle

Calculate

a. the circumference

b. the area of a circle with a radius of 6 ft.

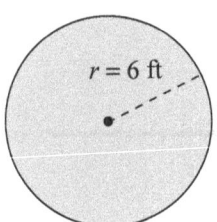

Solution

3.R.1 Formulas in Geometry 79

✏️ Now You Try It!

Use the space provided to work out the solution to the next example.

Example C Calculating the Circumference and Area of a Circle

Calculate **a.** the circumference and **b.** the area of a circle with a radius of 11 m.

Volume and Surface Area

Volume is a measure of the _____.

Volume is measured in _____

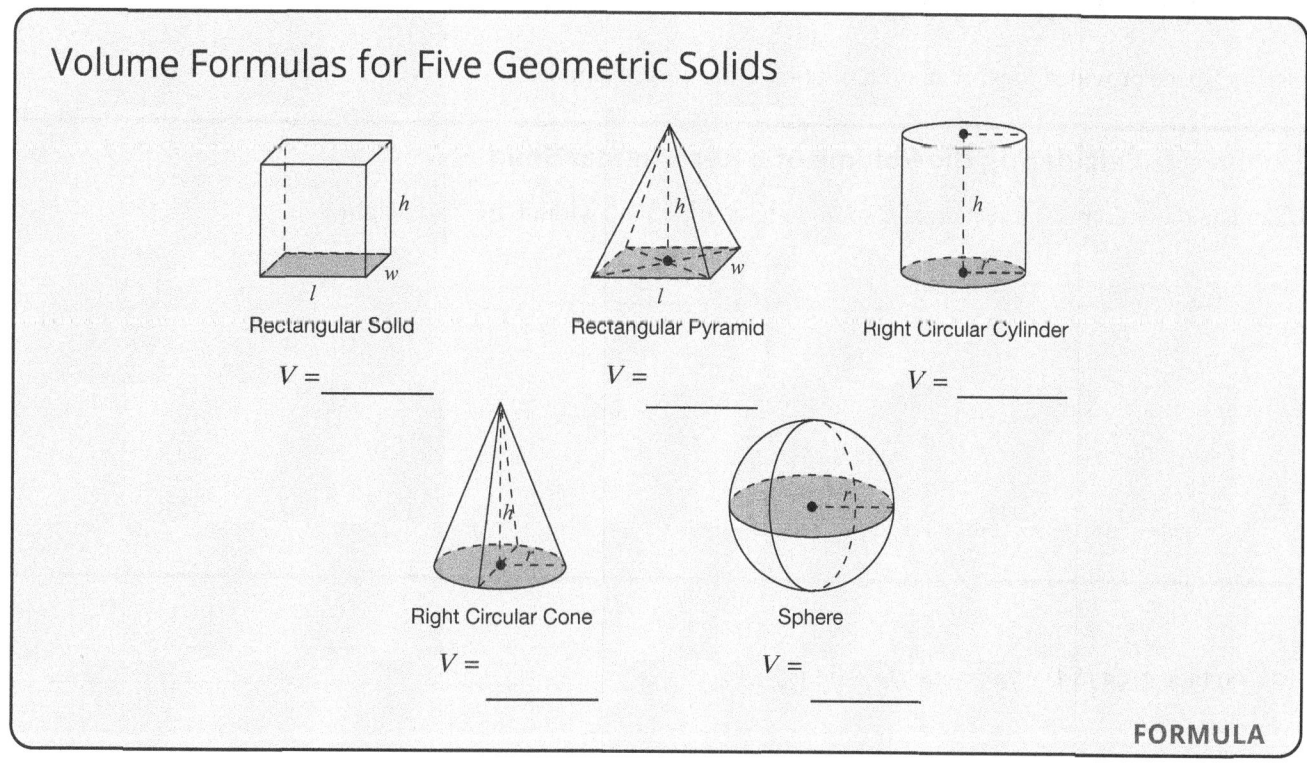

FORMULA

▶ Watch and Work

Watch the video for Example 19 in the software and follow along in the space provided.

Example 19 Calculating the Volume of a Rectangular Solid

Calculate the volume of a rectangular solid with length 8 in., width 4 in., and height 12 in.

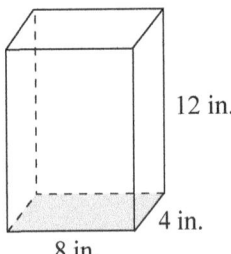

Solution

✏ Now You Try It!

Use the space provided to work out the solution to the next example.

Example D Calculating the Volume of a Rectangular Solid

Calculate the volume of a rectangular solid with length 15 in., width 6 in., and height 9 in.

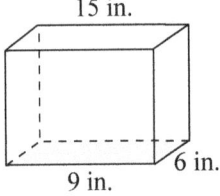

The **surface area** (*SA*) of a geometric solid is _____

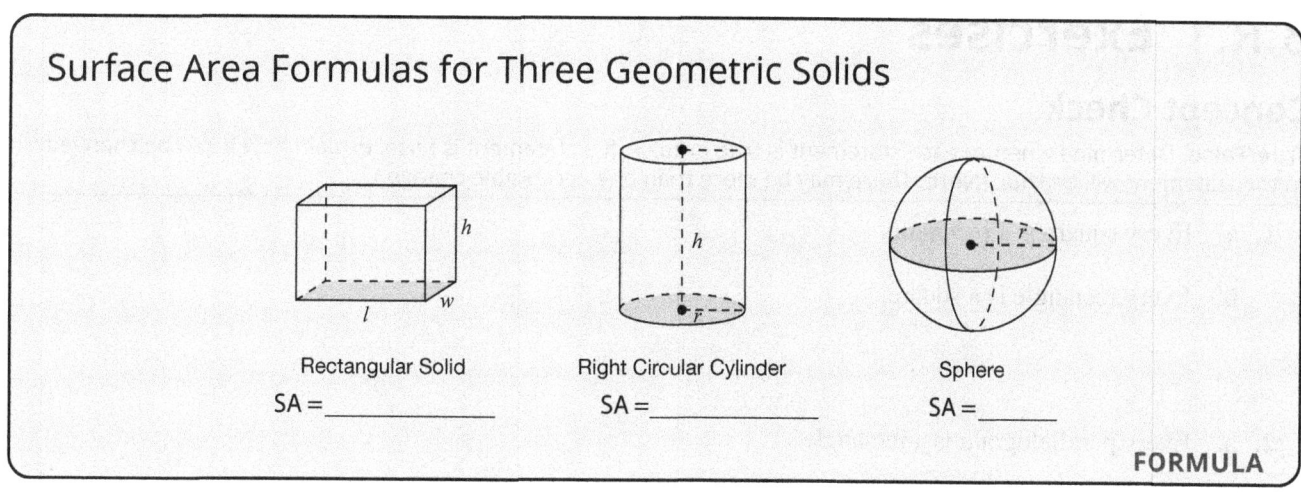

Looking Ahead

Your review of geometry will be helpful when determining properties of geometric shapes. The following example could not be answered without knowing the formula for the area of a rectangle is $A = lw$.

Example Preview

The back of Jake's property is a creek. Jake would like to enclose a rectangular garden, using the creek as one side and fencing for the other three sides. If he has 200 feet of fencing available, what is the maximum possible area of the garden?

Solution

If we let x represent the length of one side of the plot, then the dimensions of the plot are x feet by $200 - 2x$ feet. Name a new function A for area. We want to find the maximum possible value of $A(x) = x(200 - 2x)$. Simplifying produces a quadratic function as follows.

$$A(x) = x(200 - 2x) = 200x - 2x^2$$

Since $A(x) = 200x - 2x^2$ is a quadratic function with a negative leading coefficient, it represents a parabola opening downward, so its vertex will be the maximum of the function. Note the coefficients are $a = -2$, $b = 200$, and $c = 0$. The vertex is found as follows.

$$x = -\frac{b}{2a} = -\frac{200}{2(-2)} = 50$$

Substituting $x = 50$ into the function $A(x) = 200x - 2x^2$ gives the maximum value.

$$A(50) = 200(50) - 2(50)^2 = 10{,}000 - 5000 = 5000$$

Therefore, the maximum area is 5000 square feet.

3.R.1 Exercises

Concept Check

True/False. Determine whether each statement is true or false. If a statement is false, explain how it can be changed so the statement will be true. (**Note:** There may be more than one acceptable change.)

1. a. Every square is a rectangle.

   b. Every rectangle is a square.

2. a. Every parallelogram is a rectangle.

   b. Every rectangle is a parallelogram.

3. A trapezoid has only one pair of parallel lines.

4. The $(b+c)$ in the trapezoid area formula represents the sum of the lengths of the base and the corners.

5. The height of a triangle is the distance between the base and the vertex opposite the base.

6. Every radius on a circle has the same length.

7. The length of the diameter of a circle is half of the length of the radius.

8. To find the volume of a can of corn, the formula $V = \pi r^2 h$ would be used.

9. The area of the paper label on a can of peaches is an example of surface area.

10. To find the volume of a rectangular solid, the areas of each surface are added together.

3.R.1 Formulas in Geometry 83

Match each formula for perimeter to its corresponding geometric figure.

11.
a. Square   A. $P = 2l + 2w$

b. Parallelogram   B. $P = 4s$

c. Rectangle   C. $P = 2b + 2a$

d. Trapezoid   D. $P = a + b + c$

e. Triangle   E. $P = a + b + c + d$

Match each formula for volume to its corresponding geometric figure.

12.
a. Rectangular solid   A. $V = \frac{4}{3}\pi r^3$

b. Rectangular pyramid   B. $V = \frac{1}{3}\pi r^2 h$

c. Right circular cylinder C. $V = lwh$

d. Right circular cone   D. $V = \pi r^2 h$

e. Sphere   E. $V = \frac{1}{3}lwh$

Practice

Calculate the perimeter of each figure.

13. A parallelogram with sides of length 15 cm and 7 cm.

14. A square with sides of length $4\frac{1}{2}$ km.

15.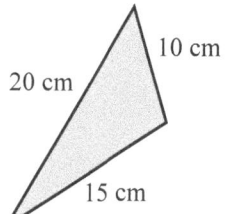

16.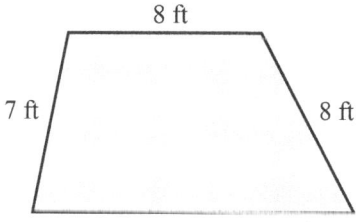

17.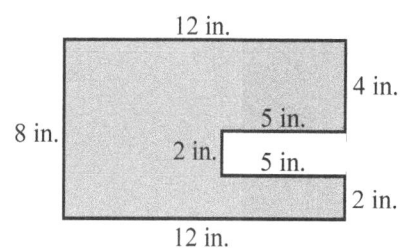

Calculate the volume of the solid. Use π ≈ 3.14

18.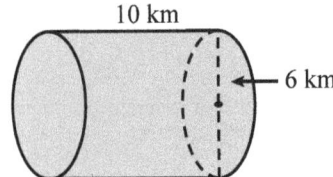

19. A rectangular solid with length 5 in., width 2 in., and height 7 in.

20. A right circular cone 3 mm high with a 2 mm radius.

Calculate the area of each figure.

21. A square with sides of length 9 ft.

22. A parallelogram with height 2.3 ft and base 11.9 ft.

24.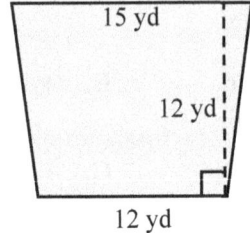

23.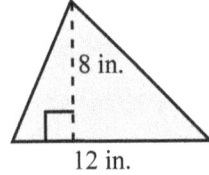

25.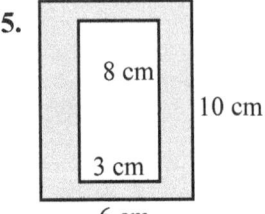

Calculate **a.** the perimeter and **b.** the area of each figure. Use π ≈ 3.14.

26.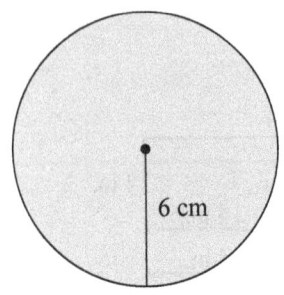

27.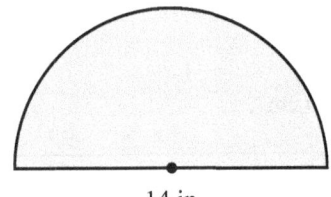

28.

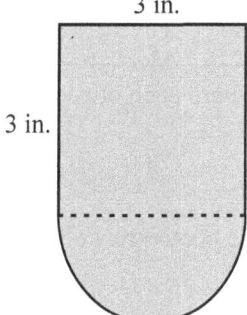

Calculate the area of the shaded portion of the figure. Use π ≈ 3.14.

29.

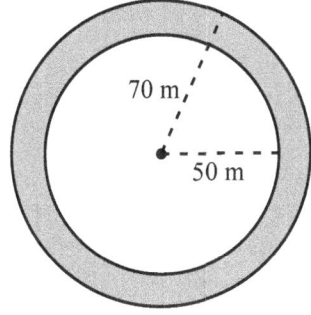

Calculate the surface area of each solid. Use π ≈ 3.14.

30.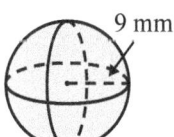

31.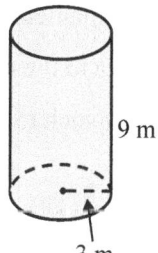

Applications

Solve. Use π ≈ 3.14.

32. **Construction:** The Pentagon near Washington, D.C., is a five-sided building where each outside wall is 921 feet.

    a. What is the perimeter of the building?

    b. If it takes a person 0.00341 minutes to walk 1 foot, how long will it take the person to walk completely around the building? Round your answer to the nearest tenth of a minute.

33. **Construction:** The main stage at a theater is in the shape of a trapezoid. The owner of the theater is planning to install a new specially designed flooring system on the stage. The stage is 12 feet wide in the front and 15 feet wide in the back. The stage is 10 feet deep. How much flooring will the manager need?

    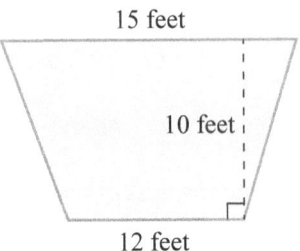

34. **Technology:** A square electronics circuit board is 18 centimeters on each side. On the center of one of the edges is an 8 by 1.5 centimeter rectangular lip for plugging in.

    a. What is the total perimeter of the circuit board, including the lip?

    b. What is the area of the circuit board?

    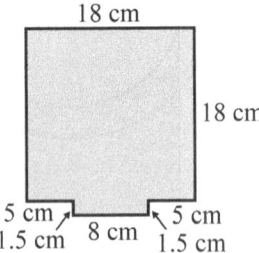

35. **Sales:** Papa Luigi's sells a 9-inch diameter pizza for $8.

    a. Determine the area of the pizza to the nearest tenth.

    b. Determine the price per square inch to the nearest cent per square inch.

36. **Geometry:** Disposable paper drinking cups like those used at water coolers are often cone-shaped. Find the volume of such a cup that is 9 cm high with a 3.2 cm radius. Express the answer to the nearest milliliter.

    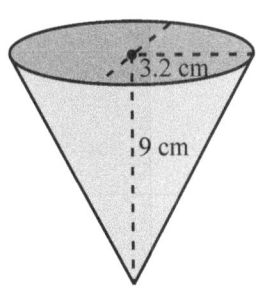

Writing & Thinking

37. List the steps and formulas you would use to find the volume of an ice cream cone (assuming the ice cream itself forms a perfect half sphere).

38. Explain what the value of $(b+c)$ represents in the formula for the area of a trapezoid.

39. Explain why $2\pi r$ is equivalent to πd.

40. Propose a method for calculating the area of a semicircle and justify your method.

3.R.2 Square Roots and the Pythagorean Theorem

↻ Making Connections

The distance formula and the standard form of the equation of a circle look very similar, because they are both derived from the Pythagorean Theorem. When solving for the distance between two points and working with circle equations, you will often find your answer includes a square root. Therefore, it is necessary to review the Pythagorean Theorem and square roots.

In this section, you will learn skills that you can apply when answering questions like these:

- Calculate the distance between the following pair of points: (5,1) and (−1,3).
- Sketch the graph of the following equation by plotting points.

$$x^2 + y^2 - 6x = 0$$

- Find the standard form of the equation for the circle with a diameter whose endpoints are (−4,−1) and (2,5).

🛠 Building Foundations

> **Terminology of Radicals**
>
> The symbol $\sqrt{}$ is called _____
>
> The number under the radical sign is called _____
>
> The complete expression, such as $\sqrt{49}$, is called _____
>
> DEFINITION

▶ Watch and Work

Watch the video for Example 2 in the software and follow along in the space provided.

Example 2 Evaluating Square Roots

Use your memory of the values in Table 2 to evaluate each expression.

a. $\sqrt{256}$

b. $\sqrt{81}$

Solution

✏️ Now You Try It!

Use the space provided to work out the solution to the next example.

Example A Evaluating Square Roots

Evaluate each expression.

a. $\sqrt{36}$

b. $\sqrt{169}$

Terms Related to Right Triangles

Right triangle: _____

Hypotenuse: _____

Leg: _____
_____

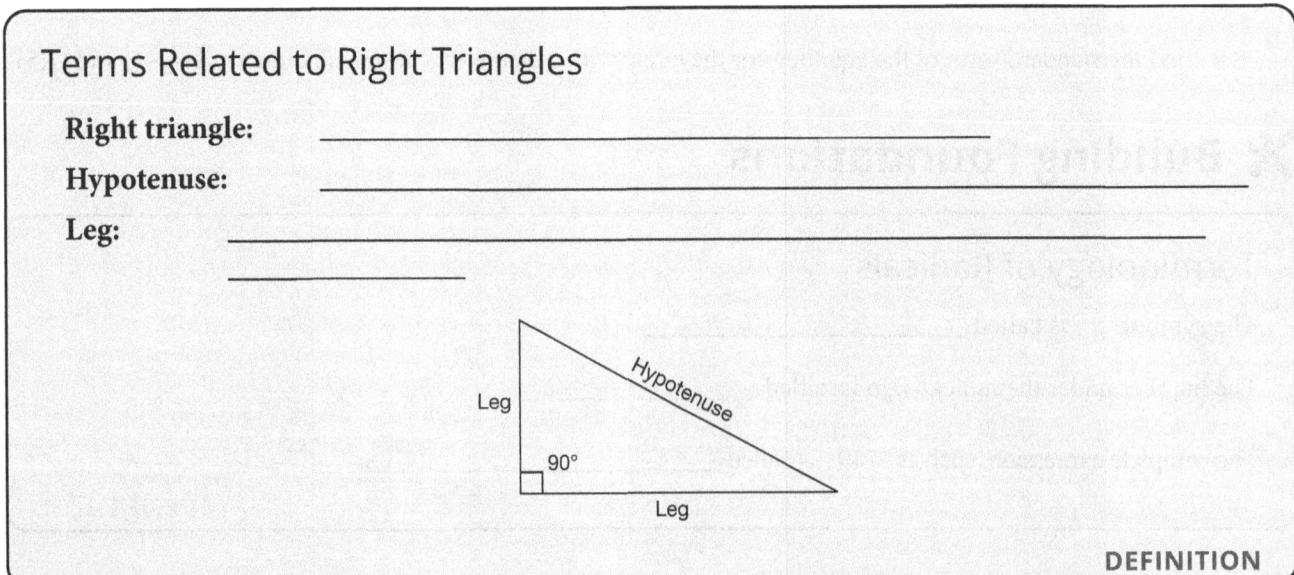

DEFINITION

The Pythagorean Theorem

In a right triangle, the _____
_____

____ = ____ + ____

THEOREM

Looking Ahead

Your review of the Pythagorean Theorem will be helpful in determining the equation for a circle given its center and a point on the circle. Recall that all the points on a circle lie at a specific distance from its center. This distance is the radius r of the circle.

Example Preview

Find the standard form of the equation for the circle with the following properties.

$$\text{Center } (11,2), \text{ passes through } (2,-10)$$

Solution

The standard form of the equation for a circle is

$$(x-h)^2 = (y-k)^2 = r^2$$

where the center is (h,k) and the radius has length r. Since every point on a circle is the same distance r from the circle's center, we can conclude that the radius r equals the distance from the center, $(11,2)$, to the point given in the problem, $(2,-10)$. We use the distance formula to find this value.

$$r = \sqrt{(x_2 - x_1)^2 + (y_2 - y_1)^2}$$
$$= \sqrt{(2-(11))^2 + (-10-(2))^2}$$
$$= \sqrt{81 + 144}$$
$$= 15$$

Now we substitute this and the values given in the problem and simplify to obtain the following equation.

$$(x-11)^2 + (y-2)^2 = (15)^2$$
$$(x-11)^2 + (y-2)^2 = 225$$

3.R.2 Exercises

Concept Check

True/False. Determine whether each statement is true or false. If a statement is false, explain how it can be changed so the statement will be true. (**Note:** There may be more than one acceptable change.)

1. 49 is a perfect square.

2. In the expression $\sqrt{81}$, the number 9 is the radicand.

3. The Pythagorean Theorem can be used to find the length of the longest side of a right triangle if the lengths of the two legs are known.

4. The Pythagorean Theorem works for any type of triangle.

Practice

Evaluate each expression.

5. $\sqrt{36}$

6. $\sqrt{225}$

Use the Pythagorean Theorem to determine whether or not each triangle is a right triangle.

7.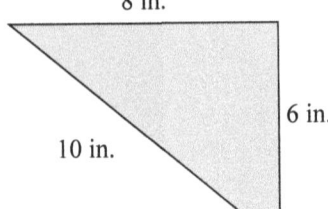

8.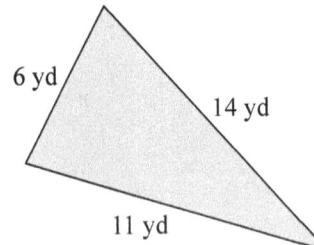

Find the hypotenuse for each right triangle accurate to the nearest hundredth.

9.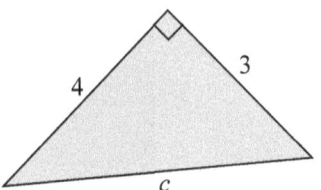

Applications

Solve.

10. **Safety:** The base of a fire engine ladder is 30 feet from a building and reaches to a third floor window 50 feet above ground level. Find the length of the ladder to the nearest hundredth of a foot.

11. **Baseball:** The shape of home plate in the game of baseball can be created by cutting off two triangular pieces at the corners of a square, as shown in the figure. If each of the triangular pieces has a hypotenuse of 12 inches and legs of equal length, what is the length of one side of the original square, to the nearest tenth of an inch?

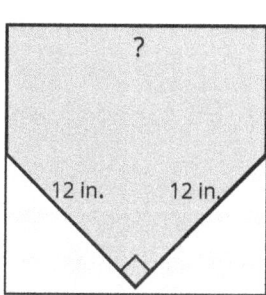

Writing & Thinking

12. Explain the connection between a perfect square and its square root. Give an example.

3.R.3 Evaluating Radicals

↻ Making Connections

The Pythagorean Theorem appears in several areas of mathematics. For example, it is needed to determine the distance between two points and to write the equation of a circle. Being able to evaluate radicals is a necessary skill to navigate such topics.

In this section, you will learn skills that you can apply when answering questions like these:

- Determine the distance between the points $(-1,-9)$ and $(5,9)$.
- Find the radius of the following circle.

$$(x+2)^2 + (y-5)^2 = 9$$

🛠 Building Foundations

Radical Terminology

The symbol $\sqrt{}$ is called _____

The number under the _____

The complete expression, such as $\sqrt{64}$ is called _____

DEFINITION

Square Root

If a is a nonnegative real number, then

_____

_____

_____

DEFINITION

Cube Root

If a is a real number, then _____

DEFINITION

▶ Watch and Work

Watch the video for Example 4 in the software and follow along in the space provided.

Example 4 Evaluating Cube Roots

a. Because $2^3 = \_$, $\sqrt[3]{8} = \_$.

b. Because $(-6)^3 = \_\_\_$, $\sqrt[3]{-216} = \_$.

---

c. Because $\left(\dfrac{1}{3}\right)^3 = \_$, $\sqrt[3]{\dfrac{1}{27}} = \_$.

✏️ Now You Try It!

Use the space provided to work out the solution to the next example.

Example A Evaluating Cube Roots

Evaluate the following radical expressions.

a. $\sqrt[3]{64}$

b. $\sqrt[3]{-125}$

c. $\sqrt[3]{\dfrac{1}{1000}}$

👀 Looking Ahead

Now that you have reviewed and mastered evaluating radicals, you can apply this skill to problems involving more complicated expressions such as the one in the following example.

Example Preview

Solve the following equation.

$$\sqrt{x^2 - 28} = 6$$

Solution

$$\sqrt{x^2 - 28} = 6$$

Begin by squaring both sides of the equation to eliminate the square root.

$$x^2 - 28 = 36$$

Now, since there is only an x^2 variable, we can "extract roots" by adding 28 to both sides and then take the square root of both sides.

$$x^2 = 64$$
$$x = \pm\sqrt{64}$$
$$x = \pm 8$$

Since both of these do check in the original equation, the correct answers are $x = 8$ and $x = -8$.

3.R.3 Exercises

Concept Check

True/False. Determine whether each statement is true or false. If a statement is false, explain how it can be changed so the statement will be true. (**Note:** There may be more than one acceptable change.)

1. If a number is squared and the principal square root of the result is found, that square root is always equal to the original number.

2. There is no real number that can be a square root of a negative number.

3.R.3   Evaluating Radicals

3. The index is the number underneath the radical sign.

4. The cube root of –27 is a real number.

Practice
Simplify the following square roots and cube roots.

5. $\sqrt{49}$

6. $\sqrt{289}$

7. $\sqrt[3]{1000}$

8. $\sqrt[3]{\dfrac{27}{64}}$

9. $\sqrt{0.04}$

Applications

Solve.

10. **Area:** The area of a square tile is 16 square inches.

    a. How long are the sides of the square tile?

    b. How many tiles would be needed for a four-foot-long and four-inch-high backsplash in a newly designed bathroom?

11. **Volume:** The volume of a child's building block is 64 cubic centimeters.

    a. Assuming the building block is a perfect cube, find the length of each side of the block.

    b. If a child stacks 5 blocks directly on top of each other, find the height of the structure that is created.

Writing & Thinking

12. Discuss, in your own words, why the square root of a negative number is not a real number.

13. Discuss, in your own words, why the cube root of a negative number is a negative number.

3.R.4 Simplifying Radicals

♻ Making Connections

When simplifying radical expressions and transforming expressions with rational exponents into equivalent radical form, you will often have to use a variety of rules. Therefore, it is important to review how to simplify radicals by finding the square root of a variable with even exponents, the square root of a variable with odd exponents, and the cube root of an algebraic expression.

In this section, you will learn skills that you can apply when answering questions like these:

- Simplify the following radical expression.

$$\frac{\sqrt{-98}}{3i\sqrt{-2}}$$

- Simplify the following radical expression.

$$\sqrt[3]{-8x^6y^9}$$

- Simplify the following radical expression.

$$\sqrt{y^4} \cdot \sqrt[6]{y^3}$$

🛠 Building Foundations

Properties of Square Roots

If a and b are **positive** real numbers, then

1. _____

2. _____

PROPERTIES

Simplest Form for Square Roots

A square root is considered to be in **simplest form** when _____
_____

DEFINITION

Square Root of x^2

If x is a real number, then _____

Note: If $x \geq 0$ is given, then _____

DEFINITION

▶ Watch and Work

Watch the video for Example 3 in the software and follow along in the space provided.

Example 3 Simplifying Square Roots with Variables

Simplify each of the following square roots. Look for perfect square factors and even powers of the variables. Assume that all variables represent positive real numbers.

a. $\sqrt{81x^4}$

b. $\sqrt{64x^5 y}$

c. $\sqrt{18a^4 b^6}$

d. $\sqrt{\dfrac{9a^{13}}{b^4}}$

Solution

3.R.4 Simplifying Radicals

✏ Now You Try It!

Use the space provided to work out the solution to the next example.

Example A Simplifying Square Roots with Variables

Simplify each of the following square roots. Assume that all variables represent positive real numbers.

a. $\sqrt{16x^8}$

b. $\sqrt{100x^3y^3}$

c. $\sqrt{12x^8y^{12}}$

d. $\sqrt{\dfrac{25z^{18}}{y^8}}$

Simplest Form for Cube Roots

A cube root is considered to be in _____

DEFINITION

3.R.4 Simplifying Radicals

🔭 Looking Ahead

In the following example, you will find the square root of an expression that contains variables raised to exponents that are not all even.

Example Preview

Simplify the following radical expression.

$$\sqrt{14y^{14}z}$$

Solution

This expression can be simplified as follows.

$$\begin{aligned}\sqrt{14y^{14}z} &= \sqrt{14} \cdot \sqrt{y^{14}} \cdot \sqrt{z} \\ &= \sqrt{14} \cdot \sqrt{z} \cdot |y^7| \\ &= \sqrt{14z} \cdot |y^7| \\ &= |y^7|\sqrt{14z}\end{aligned}$$

3.R.4 Exercises

Concept Check

True/False. Determine whether each statement is true or false. If a statement is false, explain how it can be changed so the statement will be true. (**Note:** There may be more than one acceptable change.)

1. Any variable term with an exponent of 5 has a perfect cube factor within that variable term.

2. The simplest form of a radical expression can be found by using prime factorization.

3. If x is a real number, then $\sqrt{x^2} = x$.

4. The term $7b\sqrt[3]{6c^2}$ is in simplified form.

Practice
Simplify each of the following radical expressions. Assume that all variables represent positive real numbers.

5. $\sqrt{162}$

6. $\sqrt{\dfrac{32}{49}}$

7. $\sqrt{24x^{11}y^2}$

8. $\sqrt[3]{56}$

9. $\sqrt[3]{-8x^8}$

Applications

Use the following two formulas associated with electricity

$I = \sqrt{\dfrac{P}{R}}$

$E = \sqrt{PR}$

P = power (in watts)
I = current (in amperes)
E = voltage (in volts)
R = resistance (in ohms, Ω)

10. *Electricity:* What is the current in amperes of a light bulb that produces 150 watts of power and has a 25 Ω resistance?

11. *Electricity:* If a light bulb has a resistance of 30 Ω and produces 90 watts of power, what is its current in amperes?

Writing & Thinking

12. Under what conditions is the expression $\sqrt{a}$ not a real number?

13. Explain why the expression $\sqrt[3]{y}$ is a real number regardless of whether $y > 0$, $y < 0$, or $y = 0$.

3.R.5 Introduction to the Cartesian Coordinate System

↻ Making Connections

Understanding how the Cartesian Coordinate system works is one of the most important skills you can develop as you prepare to study more advanced topics. It allows us to make important connections between geometry, trigonometry, and algebra. For example, you will be able to represent lines and circles using equations.

In this section, you will learn skills that you can apply when answering questions like these:

- Express the following equation in slope-intercept form and graph the line $2x+4y+8=0$.
- Plot the point $A(-3,3)$ on the graph.

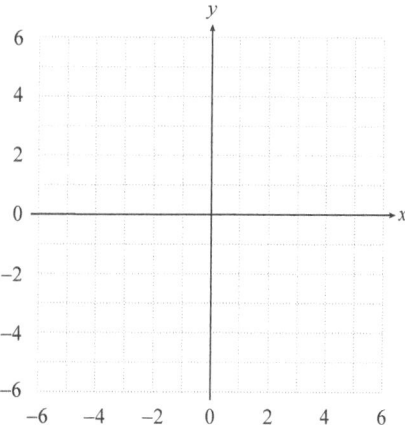

🛠 Building Foundations

Descartes based his system on a relationship between _____ in a plane and _____ _____ of real numbers.

In the ordered pair (x, y), x is called the _____ and y is called the _____ _____.

In an ordered pair of the form (x, y), the _____ is called the **independent variable** and the _____ is called the **dependent variable**.

3.R.5 Introduction to the Cartesian Coordinate System

The Cartesian coordinate system relates algebraic equations and ordered pairs to geometry. In this system, two number lines intersect at right angles and separate the plane into four _____. The **origin**, designated by the ordered pair $(0, 0)$, is _____. The horizontal number line is called the _____ or _____. The vertical number line is called the _____ or _____.

> **One-to-One Correspondence**
>
> _____
>
> _____
>
> DEFINITION

▶ Watch and Work

Watch the video for Example 4 in the software and follow along in the space provided.

Example 4 Finding Ordered Pairs

Complete the table so that each ordered pair will satisfy the equation $y = -3x + 1$.

x	y	(x, y)
0		
	4	
$\frac{1}{3}$		
3		

Solution

✏️ Now You Try It!

Use the space provided to work out the solution to the next example.

Example A Finding Ordered Pairs

Complete the table so that each ordered pair will satisfy the equation $y = -3x + 2$.

x	y	(x, y)
0		
	1	
−2		
	0	

Looking Ahead

Now that you have reviewed the main ideas around the Cartesian coordinate system, you will apply these skills to more advanced problems. For example, the next exercise involves identifying key features of a circle and then graphing the circle on the Cartesian plane.

Example Preview

Consider the following equation of a circle. Find the center (h,k), the radius r, and graph the circle.

$$(x-6)^2 + (y-5)^2 = 16$$

Solution

The standard form of the equation of a circle is $(x-h)^2 + (y-k)^2 = r^2$.

The center is $(h,k) = (6,5)$ and the radius is $r = 4$ because $r^2 = 16$.

Plot the center of the circle at $(6,5)$ and size the circle such that the points on the circle are 4 units away from the center.

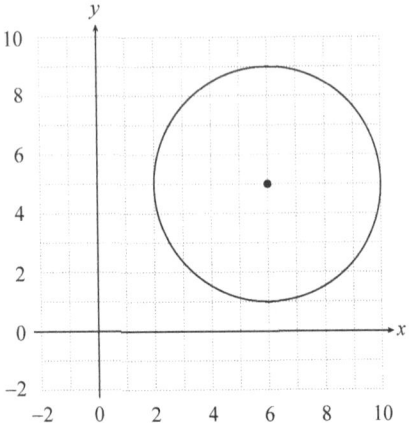

3.R.5 Exercises

True/False. Determine whether each statement is true or false. If a statement is false, explain how it can be changed so the statement will be true. (**Note:** There may be more than one acceptable change.)

1. The graph of every ordered pair that has a positive x-coordinate and a negative y-coordinate can be found in Quadrant IV.

2. To find the y-value that corresponds with $x = 2$, substitute 2 for x into the given equation and solve for y.

3. If $(-7, 3)$ is a solution of $y = 3x + 24$, then $(-7, 3)$ satisfies $y = 3x + 24$.

4. If point $A = (0, 4)$, then point A lies on the x-axis.

Practice

List the set of ordered pairs corresponding to the points on the graph.

5.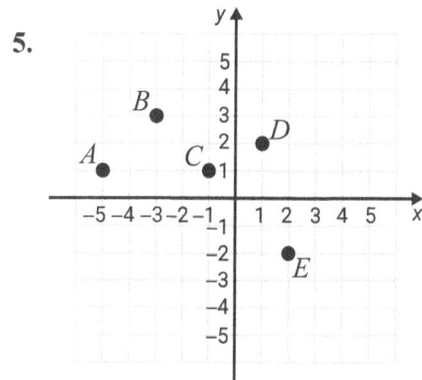

Plot each set of ordered pairs and label the points.

6. $\{A(4, -1), B(3, 2), C(0, 5), D(1, -1), E(1, 4)\}$

3.R.5 Introduction to the Cartesian Coordinate System

Determine the missing coordinate in each of the ordered pairs so that the point will satisfy the equation given.

7. $x - 2y = 2$
   a. $(0, \_\_)$
   b. $(4, \_\_)$
   c. $(\_\_, 0)$
   d. $(\_\_, 3)$

Complete the tables so that each ordered pair will satisfy the given equation. Plot the resulting sets of ordered pairs.

8. $y = 2x - 3$

x	y
0	
	-1
-2	
	3

Determine which, if any, of the ordered pairs satisfy the given equation.

9. $2x - 3y = 7$
   a. $(1, 3)$
   b. $\left(\dfrac{1}{2}, -2\right)$
   c. $\left(\dfrac{7}{2}, 0\right)$
   d. $(2, 1)$

The graph of a line is shown. List any three points on the line. (There is more than one correct answer.)

10.
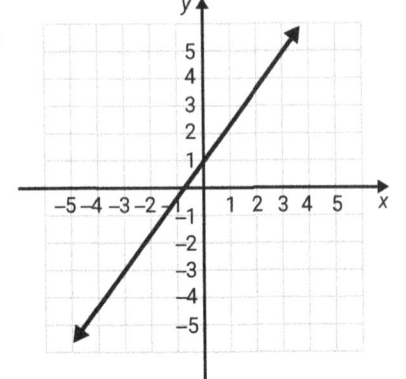

Applications

Solve.

11. **Exchange Rates:** At one point in 2017, the exchange rate from US dollars to Euros was $E = 0.85D$ where E is Euros and D is dollars.

    a. Make a table of ordered pairs for the values of D and E if D has the values $100, $200, $300, $400, and $500.

    b. Plot the points corresponding to the ordered pairs.

12. **Temperature:** Given the equation $F = \frac{9}{5}C + 32$ where C is temperature in degrees Celsius and F is the corresponding temperature in degrees Fahrenheit:

    a. Make a table of ordered pairs for the values of C and F if C has the values $-20°$, $-10°$, $-5°$, $0°$, $5°$, $10°$, and $15°$.

    b. Plot the points corresponding to the ordered pairs.

3.R.6 Solving Linear Equations: $ax + b = c$

↻ Making Connections

While straight lines are, in their essence, geometric objects, they can be represented very efficiently using linear equations. The ability to solve different types of linear equations is very helpful when learning to graph lines.

In this section, you will learn skills that you can apply when answering questions like these:

- Determine the missing coordinate in the ordered pair $(2,?)$ so that it will satisfy the following equation.

$$6x + 5y = 8$$

- For the following equation, determine the values of the missing entries in the table.

$$8x - 4y = 18$$

x		0	1	
y	0			2

✖ Building Foundations

Procedure for Solving Linear Equations that Simplify to the Form $ax + b = c$

1. Combine _____

2. Use the **addition principle of equality** and _____
   _____

3. Use the _____ and multiply both sides of the equation by the reciprocal of the coefficient of the variable (or _____
   _____ **itself**). The _____ will become +1.

4. Check your answer by _____

PROCEDURE

▶ Watch and Work

Watch the video for Example 2 in the software and follow along in the space provided.

Example 2 Solving Linear Equations of the Form $ax + b = c$

Solve the equation: $-26 = 2y - 14 - 4y$

Solution

✏ Now You Try It!

Use the space provided to work out the solution to the next example.

Example A Solving Linear Equations of the Form $ax + b = c$

Solve the equation.

$-18 = 2y - 8 - 7y$

👁 Looking Ahead

Now that you have mastered solving equations of the form $ax + b = c$, you will be able to apply these skills to problems involving the point-slope form of the equation of a line, $y = mx + b$, as in the next example.

Example Preview

Express the given equation in slope-intercept form. Then plot the point on the line with the x-coordinate $x = 4$.

$$3x - 4y - 28 = 0$$

Solution

First, solve for y to get the equation in slope-intercept form.

$$3x - 4y - 28 = 0$$
$$3x - 4y - 28 - 3x = 0 - 3x$$
$$-4y - 28 = -3x$$
$$-4y - 28 + 28 = -3x + 28$$
$$-4y = -3x + 28$$
$$\frac{-4y}{-4} = \frac{-3x}{-4} + \frac{28}{-4}$$
$$y = \frac{3}{4}x - 7$$

Now, substitute the given value of x into the equation and simplify.

$$y = \frac{3}{4}x - 7$$
$$y = \frac{3}{4}(4) - 7$$
$$y = 3 - 7$$
$$y = -4$$

The corresponding point is $(4, -4)$.

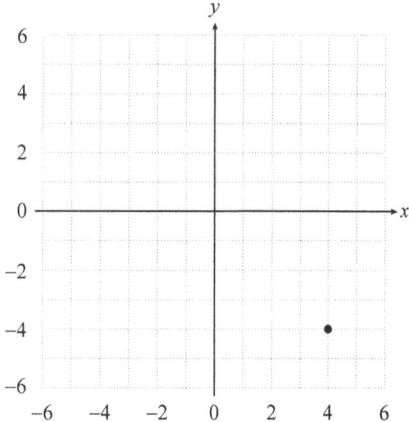

3.R.6 Exercises

Concept Check

True/False. Determine whether each statement is true or false. If a statement is false, explain how it can be changed so the statement will be true. (**Note:** There may be more than one acceptable change.)

1. If an equation of the form $ax + b = c$ uses decimal or fractional coefficients, the addition and multiplication principles of equality cannot be used.

2. The first step in solving $2x + 3 = 9$ is to add 3 to both sides.

3. To solve an equation that has been simplified to $4x = 12$, you need to multiply both sides by $\frac{1}{4}$, or divide both sides by 4.

4. When solving a linear equation with decimal coefficients, one approach is to multiply both sides in such a way to give integer coefficients before solving.

Practice

Solve each equation.

5. $3x + 11 = 2$

6. $-5x + 2.9 = 3.5$

7. $\dfrac{2}{5} - \dfrac{1}{2}x = \dfrac{7}{4}$

8. $\dfrac{y}{3} - \dfrac{2}{3} = 7$

Applications

Solve.

9. *Music:* The tickets for a concert featuring the new hit band, Flying Sailor, sold out in 2.5 hours. If there were 35,000 tickets sold, solve the equation $35,000 - 2.5x = 0$ to find the number of tickets sold per hour.

10. *Movies:* All snacks (candy, popcorn, and soda) cost $3.50 each at the local movie theater. Admission tickets cost $7.50 each. After a long week, Carlos treats himself to a night at the movies. His movie night budget is $25 and he spends all his movie money. Solve the equation $3.50x + $7.50 = $25.00 to determine how many snacks Carlos can buy.

Writing & Thinking

11. Find the error(s) made in solving each equation and give the correct solution.

    a. $\frac{1}{3}x + 4 = 9$

    $3 \cdot \frac{1}{3}x + 4 = 3 \cdot 9$

    $x + 4 = 27$

    $x + 4 - 4 = 27 - 4$

    $x = 23$

    b. $5x + 3 = 11$

    $(5x - 3) + (3 - 3) = 11 - 3$

    $2x + 0 = 8$

    $\frac{2x}{2} = \frac{8}{2}$

    $x = 4$

3.R.7 Solving Linear Equations: $ax + b = cx + d$

↻ Making Connections

Solving linear equations involving several steps is a fundamental skill necessary to describe lines on the coordinate plane. It allows us to plot points accurately, determine the slope, and determine the intercepts for a line.

In this section, you will learn skills that you can apply when answering questions like these:

- Find the *x*- and *y*-intercepts of the following equations.

$$3y + 2x = 4y + 2$$

$$3x + 7y = -21$$

- Find the slope and *y*-intercept of the following equation.

$$2x - 2y = -4$$

✖ Building Foundations

Procedure for Solving Linear Equations that Simplify to the Form $ax + b = cx + d$

1. Simplify each side of the equation by _____

   _____

2. Use the **addition principle of equality** and _____

   _____

3. Use the **multiplication** (or **division**) **principle of equality** and _____

   _____ of the coefficient of the variable (**or divide** _____

   _____ **itself**). The coefficient of the variable _____

4. Check your answer by _____

PROCEDURE

Type of Equation	Number of Solutions
conditional	_____
identity	_____
contradiction	_____

Table 1

▶ Watch and Work

Watch the video for Example 9 in the software and follow along in the space provided.

Example 9 Determining Types of Equations

Determine whether the equation $3(x-25)+3x=6(x+10)$ is a conditional equation, an identity, or a contradiction.

Solution

✏ Now You Try It!

Use the space provided to work out the solution to the next example.

Example A Determining Types of Equations

Determine whether the equation $-3(x-4) = 12 - 2x - x$ is a conditional equation, an identity, or a contradiction.

Looking Ahead

Now that you have reviewed solving linear equation of the form $ax+b=cx+d$, you will be able to apply your skills to solving multi-step problems involving multiple equations like the following example.

Example Preview

Express the following equations in slope-intercept form and determine if the two lines are perpendicular.

$$8-(3y+4x)=6(x-y) \quad \text{and} \quad 3y+2=8+10x$$

Solution

$$8-(3y+4x)=6(x-y) \qquad 3y+2=8+10x$$
$$8-3y-4x=6x-6y \qquad 3y=10x+8-2$$
$$-3y+6y=6x+4x-8 \qquad 3y=10x+6$$
$$3y=10x-8 \qquad y=\frac{10}{3}x+2$$
$$y=\frac{10}{3}x-\frac{8}{3}$$

These two lines have slopes of $\frac{10}{3}$ and $\frac{10}{3}$, respectively. Since both of these lines have the same slope, these two lines are parallel and not perpendicular.

3.R.7 Exercises

Concept Check

True/False. Determine whether each statement is true or false. If a statement is false, explain how it can be changed so the statement will be true. (**Note:** There may be more than one acceptable change.)

1. Every linear equation has exactly one solution.

2. If a linear equation simplifies to a statement that is always true, then the original equation is called an identity.

3. If an equation has no solution, it is called an identity.

4. The most general form of a linear equation is $ax+b=cx+d$.

Practice

Solve each equation.

5. $3x + 2 = x - 8$

6. $2(z+1) = 3z + 3$

7. $x - 0.1x + 0.8 = 0.2x + 0.1$

8. $0.6x - 22.9 = 1.5x - 18.4$

Determine whether each equation is a conditional equation, an identity, or a contradiction.

9. $-2x + 13 = -2(x - 7)$

10. $3x + 9 = -3(x - 3) + 6x$

Applications

Solve.

11. **Event Planning:** Caitlyn and Steve are planning their wedding reception and must decide between two catering halls. The first site, A Wedding Space, rents for $800 for one day and charges $50 per person for dinner. The second venue, A Wedding Place, costs $1000 to rent for one day and charges $40 per person for the same dinner. Solve the equation $800 + 50x = 1000 + 40x$ to determine how many guests they can invite so that the cost they pay will be the same at both wedding catering halls.

12. **Personal Finance:** The value of a new car depreciates at a rate of about $250 per month. Suppose a car originally costs $30,000. The car was bought with a $1000 down payment and a loan with 0% financing for 60 months with payments of $200 a month. Solve the equation $30{,}000 - 250t = 29{,}000 - 200t$ to determine how many months it will take for the value of the vehicle to equal the amount owed on the loan?

Writing & Thinking

13. Answer each question.

    a. Simplify the expression $3(x+5)+2(x-7)$.

    b. Solve the equation $3(x+5)+2(x-7)=31$.

    c. How are the methods you used to answer questions **a.** and **b.** similar? How are they different?

3.R.7 Solving Linear Equations: $ax + b = cx + d$

3.R.8 Solving Linear Inequalities in One Variable

↻ Making Connections

The ability to solve linear inequalities in one variable allows us to compare different quantities and represent the solution graphically, algebraically, and with interval notation. This ability will be helpful when graphing inequalities in two variables.

In this section, you will learn skills that you can apply when answering questions like these:

- Graph the solution set of the following linear inequality.

$$y - 5x < 5$$

- Rewrite the given inequality as two linear inequalities and graph the solution set.

$$|2x - 4y| \geq 7$$

🛠 Building Foundations

The set of all real numbers between a and b is called an _____.

Type of Interval	Algebraic Notation	Interval Notation	Graph
Open Interval	_____	(a, b)	
Closed Interval	$a \leq x \leq b$	_____	
_____	$\begin{cases} a \leq x < b \\ a < x \leq b \end{cases}$	$[a, b)$ $(a, b]$	

3.R.8 Solving Linear Inequalities in One Variable

Open Interval	$\begin{cases} x > a \\ x < b \end{cases}$	_____	
Half-open Interval	_____	$[a, \infty)$ $(-\infty, b]$	

Table 1

In an **open interval**, _____.

3. In a **closed interval**, _____.

4. In a **half-open interval**, _____.

5. **Linear inequalities** are inequalities that _____.

6. A **solution** to an inequality is any number that _____
_____.

Addition Principle for Solving Linear Inequalities

If A and B are algebraic expressions and C is a real number, then _____

_____

and

_____

_____

(If a real number is added to both sides of an inequality, the new inequality is _____
_____.)

PROPERTIES

Multiplication Principle for Solving Linear Inequalities

If A and B are algebraic expressions and C is a positive real number, then _____

_____

and

_____

_____

If A and B are algebraic expressions and C is a negative real number, then _____

_____

and

_____

_____

(In other words, if both sides are multiplied by a _____

_____.)

PROPERTIES

Steps for Solving Linear Inequalities

1. Combine _____

2. Use the addition principle of inequality to _____
_____

3. Use the multiplication (or division) principle of inequality to multiply (or divide) both sides by the coefficient of the variable so that _____
   If this coefficient is negative, _____

4. A quick (and generally satisfactory) check is to _____
_____
   If the statement is false, _____

PROCEDURE

▶ Watch and Work

Watch the video for Example 9 in the software and follow along in the space provided.

Example 9 Solving Linear Inequalities

Solve the inequality $6x + 5 \leq -1$ and graph the solution set. Write the solution set using interval notation.

Solution

✏ Now You Try It!

Use the space provided to work out the solution to the next example.

Example A Solving Linear Inequalities

Solve the inequality $4x + 8 > -16$ and graph the solution set. Write the solution set using interval notation.

Looking Ahead

Now that you have reviewed the main concepts involved in solving linear equalities, you can use those skills to graph linear inequalities in two variables. The following example requires you to solve the inequality for the variable y before graphing.

Example Preview

Graph the solution set of the following linear inequality.

$$6x + 5 < -6y + 5$$

Solution

First, isolate y on the left-hand side of the inequality.

$$6x + 5 < -6y + 5$$
$$6y + 6x + 5 < 5$$
$$6y + 6x < 0$$
$$6y < -6x$$
$$y < -x$$

Graph the corresponding equation $y = -x$ using a dashed line because of the strict inequality. Now, choose a test point from one of the two half-planes defined by the boundary line $y = -x$, and substitute the coordinates into the inequality. For example, the point $(2,2)$ lies in the half-plane "above" the boundary line.

$$y < -x$$
$$(2) < -(2)$$
$$2 < -2$$

This statement is *false*, so we shade the half-plane "below" the boundary line.

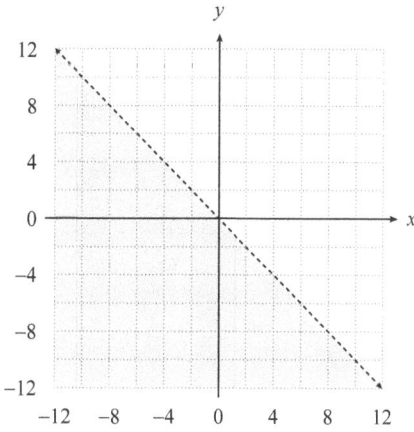

3.R.8 Exercises

Concept Check

True/False. Determine whether each statement is true or false. If a statement is false, explain how it can be changed so the statement will be true. (**Note:** There may be more than one acceptable change.)

1. If only one endpoint is included in an interval, it is called a half-open interval.

2. When both sides of a linear inequality are multiplied by a negative constant, the sense of the inequality should stay the same.

3. To check the solution set of a linear inequality, every solution in the solution set must be checked in the original inequality.

4. The infinity symbol ∞ does not represent a specific number.

Practice

Graph each interval on a real number line and tell what type of interval it is.

5. $x \leq -3$

6. $-1.5 \leq x < 3.2$

Solve each inequality and graph the solution set. Write each solution set using interval notation.

7. $x + 1 > 5$

8. $-2x \geq 6$

9. $4x - 7 \geq 9$

10. $5x + 6 \geq 2x - 2$

Applications

Solve.

11. *Test Scores:* A statistics student has grades of 82, 95, 93, and 78 on four hour-long exams. He must average 90 or higher to receive an A for the course. What scores can he receive on the final exam and earn an A if:

    a. The final is equivalent to a single hour-long exam (100 points maximum)?

    b. The final is equivalent to two hourly exams (200 points maximum)?

12. *Postage:* Allison is going to the post office to buy 34¢ stamps and 3¢ adjustment stamps. Since the current postage rate is 49¢, she will need 5 times as many 3¢ adjustment stamps as 34¢ stamps. If she has $12.25 to spend, what is the largest number of 34¢ stamps she can buy?

Writing & Thinking

13. **a.** Write a list of three situations where inequalities might be used in daily life.

    b. Illustrate theses situations with algebraic inequalities and appropriate numbers.

3.R.9 Solving Radical Equations

↻ Making Connections

The ability to solve equations involving radicals is a necessary skill when working with many of the concepts that you will encounter in precalculus. For example, working with the equation of a circle, the distance formula, or quadratic functions might require you to solve an equation involving a square root.

In this section, you will learn skills that you can apply when answering questions like these:

- Determine the values of the missing entries for the following equation.

$$y = x^2 - 10x + 25$$

x		4		8	3
y	0		1		

- Find the perimeter of the triangle whose vertices are $(-1, 2)$, $(-1, 6)$, and $(-4, -1)$.

🛠 Building Foundations

> **Method for Solving Equations with Radicals**
> 1. Isolate one of the radicals on _____
> 2. Raise both sides of the equation to _____
> 3. If the equation still contains a radical, _____
> 4. Solve the equation after _____
> 5. Be sure to check all possible solutions in _____
>   _____
>
> PROCEDURE

▶ Watch and Work

Watch the video for Example 6 in the software and follow along in the space provided.

Example 6 Solving Equations with Two Radicals

Solve the equation: $\sqrt{x+4} = \sqrt{3x-2}$

Solution

✏ Now You Try It!

Use the space provided to work out the solution to the next example.

Example A Solving Equations with Two Radicals

Solve the equation: $\sqrt{2x+5} = \sqrt{x+7}$

👀 Looking Ahead

Using the Pythagorean Theorem, our knowledge of the Cartesian Coordinate system, and our knowledge of geometric concepts, we can apply what we've learned to help us solve geometry problems involving radicals. The skills you have reviewed in this section can be applied when solving perimeter problems, as shown in the following example.

Example Preview

Find the perimeter of the triangle whose vertices are $(-3, 5)$, $(-3, 2)$, and $(-5, -3)$.

Solution

To find the perimeter of the given triangle, we should calculate the distances between each pair of vertices, and then add them up.

The vertices $(-3, 5)$ and $(-3, 2)$ have the same x-coordinate of $x = -3$, so they lie on the same vertical line. Therefore, we can determine the distance between these points without using the distance formula, just by finding the difference between the y-coordinates of the points

$$d_1 = |2 - 5| = 3$$

Both the x- and y-coordinates of the vertices $(-3, 5)$ and $(-5, -3)$ are different, so we have to use the distance formula to find the distance between these points.

$$\begin{aligned} d_2 &= \sqrt{[-5-(-3)]^2 + (-3-5)^2} \\ &= \sqrt{4 + 64} \\ &= \sqrt{68} \\ &= 2\sqrt{17} \end{aligned}$$

Similarly, we need to use the distance formula to calculate the distance between the vertices $(-3, 2)$ and $(-5, -3)$.

$$\begin{aligned} d_3 &= \sqrt{[-5-(-3)]^2 + (-3-2)^2} \\ &= \sqrt{4 + 25} \\ &= \sqrt{29} \end{aligned}$$

Finally, we add up the obtained distances and find the perimeter of the given triangle.

$$P = d_1 + d_2 + d_3 = 3 + 2\sqrt{17} + \sqrt{29}$$

3.R.9 Exercises

Concept Check

True/False. Determine whether each statement is true or false. If a statement is false, explain how it can be changed so the statement will be true. (**Note:** There may be more than one acceptable change.)

1. All radical equations will have two solutions.

2. If no true statements result when all possible solutions are checked, then there is no solution.

3. When solving equations with radicals, you should only have to raise both sides of the equation to a power one time.

4. A radical expression set equal to a negative value, such as $\sqrt{x+2} = -4$, has no real solution.

Practice

Solve the following equations. Be sure to check your answers in the original equation.

5. $\sqrt{8x+1} = 5$

6. $5 + \sqrt{x+5} - 2x = 0$

7. $\sqrt{2x-5} = \sqrt{3x-9}$

8. $\sqrt{x} + \sqrt{x-3} = 3$

Applications

Solve.

9. **Landscaping:** A landscaper is designing a pond in the shape of a right triangle that has a square flower patch along each edge. She knows two of the flower patches will have side lengths of 5 ft and 13 ft and that the remaining flower patch must have a side length a which satisfies the equation $13 = \sqrt{a^2 + 5^2}$. What is the side length of the remaining flower patch?

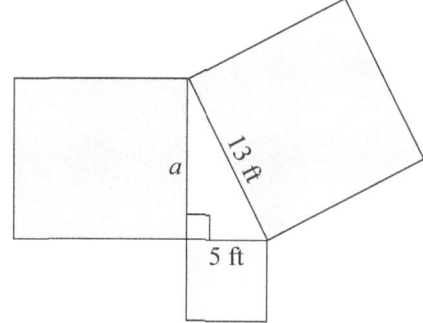

CHAPTER 4.R

Review Concepts

for *Relations, Functions, and Their Graphs*

4.R.1 Introduction to Functions and Function Notation

4.R.2 Translating English Phrases and Algebraic Expressions

4.R.3 Applications: Number Problems and Consecutive Integers

4.R.4 Greatest Common Factor (GCF) and Factoring by Grouping

4.R.5 Factoring Trinomials: $x^2 + bx + c$

4.R.6 Factoring Trinomials: $ax^2 + bx + c$

4.R.7 Review of Factoring Techniques

4.R.8 Solving Quadratic Equations by Factoring

4.R.9 Multiplication and Division with Complex Numbers

4.R.10 Quadratic Equations: The Quadratic Formula

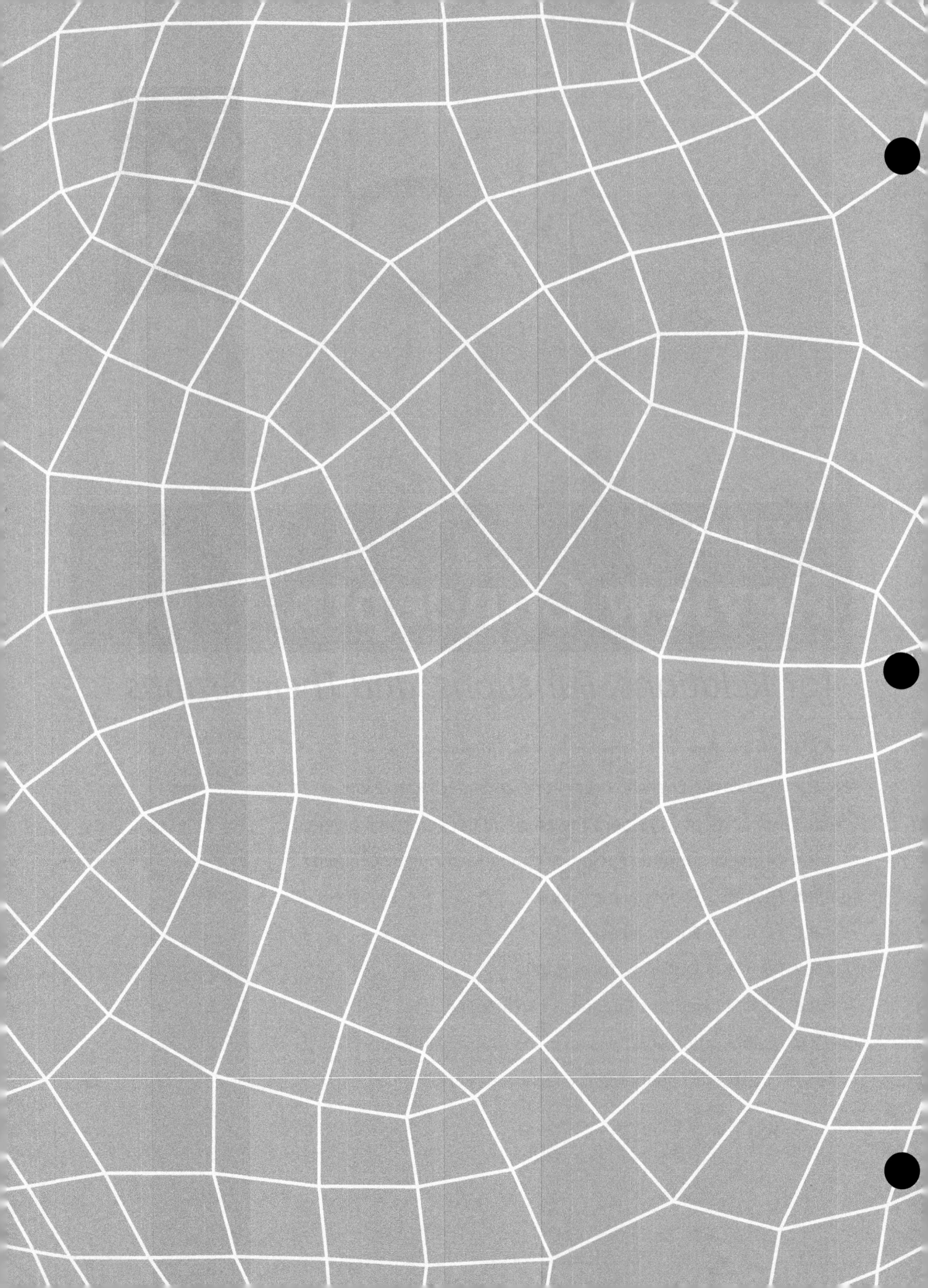

4.R.1 Introduction to Functions and Function Notation

↻ Making Connections

One of the most important concepts in all of mathematics is that of a function. Functions allows us to model the relationship between two or more quantities. For example, we can describe the position of an object as a function of time or the revenue of a company as a function of the price for their products. Understanding how to use proper notation, determine domain and range, and graph a function accurately are necessary skills to study more advanced topics.

In this section, you will learn skills that you can apply when answering questions like these:

- Determine the implied domain of the following function.

$$h(x) = \frac{5x}{x^2 - 3x - 10}$$

- Identify the domain and range of the following function.

$$h : \mathbb{N} \to \mathbb{N} \text{ by } h(x) = 5x + 9$$

- The length that a hanging spring stretches varies directly as the weight placed at the end of the spring. If a weight of 34 pounds stretches a certain spring 6 inches, how far will the spring stretch if the weight is increased to 50 pounds?

🛠 Building Foundations

> **Relation, Domain, and Range**
>
> A **relation** is a set of _____
>
> The **domain, D,** of a relation is the set of _____
>
> The **range, R,** of a relation is the set of _____
>
> DEFINITION

4.R.1 Introduction to Functions and Function Notation

Functions

A **function** is a relation in which _____

_____

DEFINITION

Vertical Line Test

If any vertical line intersects the graph of a relation at more than one point, then the relation is

_____

PROCEDURE

Linear Function

A linear function is a function represented by an equation of the form

_____

The domain of a linear function is _____

_____

DEFINITION

In function notation, instead of writing y, write _____, read "f of x."

▶ Watch and Work

Watch the video for Example 6 in the software and follow along in the space provided.

Example 6 Evaluating Functions

For the function $g(x) = 4x + 5$, find:

a. $g(2)$
b. $g(-1)$
c. $g(0)$

Solution

✏ Now You Try It!

Use the space provided to work out the solution to the next example.

Example A Evaluating Functions

For the function $g(x) = 3x - 2$, find:

a. $g(3)$
b. $g(-2)$
c. $g(0)$

4.R.1 Introduction to Functions and Function Notation

🔭 Looking Ahead

Now that you have reviewed the main ideas related to the study of functions, you will be able to tackle more advanced topics and model real-world situations using functions. In the following example, you will see how we can build a mathematical model for a person's Body Mass Index (BMI) using real data.

Example Preview

Find a model for the Body Mass Index (BMI) of a person, give that BMI varies directly as the person's weight in pounds and inversely as the square of the person's height in inches. If a 6 ft tall person weighing 143 pounds has a BMI of 19.39, how much weight would a 5 ft 4 in. tall person weighing 199 pounds need to gain or lose to have a BMI of 20?

Solution

If we let w be the weight and h be the height of the person, then the model for the BMI is

$$\text{BMI} = k\frac{w}{h^2}$$

where k is a proportionality constant. To find k, we substitute the given values into the model and solve. Note that we must convert 6 feet to 72 inches before substituting.

$$19.39 = k\frac{143}{72^2}$$

$$k = \frac{19.39 \cdot 72^2}{143} \approx 703$$

Therefore, we get the following model.

$$\text{BMI} = 703\frac{w}{h^2}$$

Now we can substitute the values of the BMI and the height to find the person's desired weight. We convert 5 ft 4 in. to 64 inches and solve for w as follows.

$$20 = 703\frac{w}{64^2}$$

$$w = \frac{20 \cdot 64^2}{703} \approx 116.5 \text{ pounds}$$

Thus, the person should lose approximately $119 - 116.5 = 2.5$ lb to have a BMI of 20.

4.R.1 Exercises

Concept Check

True/False. Determine whether each statement is true or false. If a statement is false, explain how it can be changed so the statement will be true. (**Note:** There may be more than one acceptable change.)

1. If the domain of a linear function is not explicitly stated, the implied domain is the set of all values of x that produce real values for y.

2. A relation is a function in which each domain element has exactly one corresponding range element.

3. In a function, the range elements can have more than one corresponding domain element.

4. If $s = \{(1, -6), (3, 5), (4, 0), (1, 2)\}$, then s is a function.

Practice

List the sets of ordered pairs that correspond to the points. State the domain and range and indicate if the relation is a function.

5.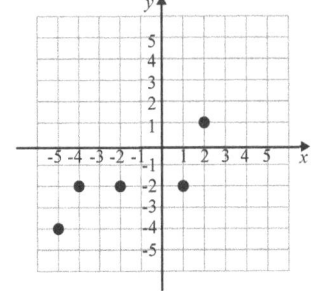

Graph the relation. State the domain and range and indicate which of the relation is a function.

6. $h = \{(1, -5), (2, -3), (-1, -3), (0, 2), (4, 3)\}$

4.R.1 Introduction to Functions and Function Notation

Use the vertical line test to determine whether the graph represents a function. State the domain and range using interval notation.

7.
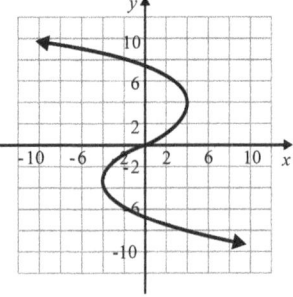

State the domain of the function.

8. $h(x) = \dfrac{7}{3x}$

Find the values of the function as indicated.

9. $F(x) = 6x^2 - 10$
   a. $F(0)$
   b. $F(-4)$
   c. $F(4)$

Applications

Solve.

10. ***Nursing:*** A nurse hangs a 1000-milliliter IV bag which is set to drip at 120 milliliters per hour. Create a model of this situation to represent the amount of IV solution left in the bag after x hours.

   a. The y-intercept is the amount of IV solution in the bag initially (time = 0). What is the y-intercept?

   b. The slope is equal to the rate that the IV solution is dispensed per hour. What is the slope? (**Hint:** Consider whether the amount of IV solution in the bag is increasing or decreasing and how this would affect the slope.)

   c. Write an equation in slope-intercept form to model this situation.

   d. Write the equation from Part **c.** using function notation.

   e. State the domain and range of the function.

   f. State any additional restrictions that should be made on the domain for it to make sense in the context of this problem.

   g. How much IV solution is left in the bag after 5 hours?

4.R.2 Translating English Phrases and Algebraic Expressions

↻ Making Connections

You will often be asked to solve mathematical problems that are presented in words. When solving these problems, it is necessary that you be able to determine what is being asked and translate the wording into something that is mathematically manageable; like an expression, function, or equation. To do so, it is imperative that you know the relevant vocabulary and the various ways in which mathematical operations can be described. Reviewing how to translate English phrases and algebraic expressions will prepare you for just such an event.

In this section, you will learn skills that you can apply when answering questions like these:

- Among all the pairs of numbers with a sum of 10, find the pair whose product is maximum.
- Among all the pairs of numbers (x,y) such that $2x + y = 20$, find the pair for which the sum of the squares is minimum.

🛠 Building Foundations

Key Words To Look For When Translating Phrases

Addition	Subtraction	Multiplication	Division	Exponent (Powers)

Division and subtraction are done with the values in the _____ that they are given in the problem.

An **ambiguous phrase** is one whose meaning is _____

_____

4.R.2 Translating English Phrases and Algebraic Expressions

▶ Watch and Work

Watch the video for Example 3 in the software and follow along in the space provided.

Example 3 Translating Algebraic Expressions into English Phrases

Change each algebraic expression into an equivalent English phrase. In each case translate the variable as "a number."

a. $5x$

b. $2n + 8$

c. $3(a - 2)$

Solution

✏️ Now You Try It!

Use the space provided to work out the solution to the next example.

Example A Identifying Properties of Addition and Multiplication

Change each algebraic expression into an equivalent English phrase.

a. $10x$

b. $4a + 7$

c. $7(n - 5)$

👀 Looking Ahead

The following example requires you to use the skills you reviewed on translating phrases into algebraic expressions. You will learn later that the minimum of a quadratic function occurs at the vertex if its graph is a parabola that opens upward.

Example Preview

Among all pairs of numbers (x,y) such that $2x + y = 16$, find the pair for which the sum of squares, $x^2 + y^2$, is minimum.

Solution

We need to write the sum of squares as a function of one variable. Solve the given linear equation for y in terms of x.

$$2x + y = 16$$

$$y = 16 - 2x$$

The sum of squares can now be written as $S(x) = x^2 + (16 - 2x)^2$

If we expand the formula for S, the result is the following quadratic function.

$$S(x) = 5x^2 - 64x + 256$$

Since the coefficient of the x^2 term is positive, this is a parabola opening upward. So, the minimum point of the graph of S is located at the vertex.

The vertex occurs at $x = -\dfrac{b}{2a}$, where a and b come from the form of a quadratic function $f(x) = ax^2 + bx + c$.

Since the quadratic function is written as $S(x) = 5x^2 - 64x + 256$, we have $a = 5$ and $b = (-64)$. The value of x can be found as follows:

$$x = -\frac{b}{2a} = -\frac{(-64)}{2(5)} = \frac{32}{5}$$

This value can be substituted back into the equation $y = 16 - 2x$ to find the value of y.

$$y = 16 - 2x = 16 - 2\left(\frac{32}{5}\right) = \frac{16}{5}$$

Therefore, the two numbers are $\dfrac{32}{5}$ and $\dfrac{16}{5}$.

4.R.2 Exercises

Concept Check

True/False. Determine whether each statement is true or false. If a statement is false, explain how it can be changed so the statement will be true. (**Note:** There may be more than one acceptable change.)

1. The order in which the values are given is particularly important when working with subtraction and division problems.

2. "More than" and "increased by" are key phrases specifying the operation of subtraction.

3. Division is indicated by the phrase "five less than a number."

4. Key phrases for parentheses can be used to limit ambiguity in English phrases.

Practice

Write the algebraic expressions described by the English phrases. Choose your own variable.

5. six added to a number

6. twenty decreased by the product of four and a number

7. eighteen less than the quotient of a number and two

Translate each pair of English phrases into algebraic expressions. Notice the differences between the algebraic expressions and the corresponding English phrases.

8. a. six less than a number

   b. six less a number

9. a. six less than four times a number

   b. six less four times a number

Write the algebraic expression described by the English phrase using the given variables.

10. the cost of purchasing a fishing rod and reel if the rod costs x dollars and the reel costs $8 more than twice the cost of the rod

Translate each algebraic expression into an equivalent English phrase. (There may be more than one correct translation.)

11. $-9x$

12. $\dfrac{9}{x+3}$

Writing & Thinking

13. Explain why translating addition and multiplication problems from English into algebra may be easier than changing subtraction or division problems. (Consider the properties previously studied.)

14. Explain the difference between $5(n+3)$ and $5n+3$ when converting from algebra to English.

4.R.3 Applications: Number Problems and Consecutive Integers

🔄 Making Connections

In mathematics and in real life you will often have to solve applications or word problems. Therefore, it is important to review how to write expressions and equations that represent given phrases or sentences in a word problem.

In this section, you will learn skills that you can apply when answering questions like these:

- The number a is 3 more than the number b. If the product of these numbers is 28, find the numbers.
- Find three consecutive odd integers whose sum is 165.
- Kathy buys last year's best-selling novel, in hardcover, for $15.05. This is a 30% discount from the original price. What was the original price?

🛠 Building Foundations

Consecutive Integers

Integers are **consecutive** if each is _____

Three consecutive integers can be represented as

_____

where n is an integer.

DEFINITION

Consecutive Even Integers

Even integers are **consecutive** if each is _____

Three consecutive even integers can be represented as

_____

where n is an **even** integer.

DEFINITION

4.R.3 Applications: Number Problems and Consecutive Integers

> **Consecutive Odd Integers**
>
> Odd integers are **consecutive** if each is _____
>
> Three consecutive odd integers can be represented as
>
> _____
>
> where n is an **odd** integer.
>
> **DEFINITION**

▶ Watch and Work

Watch the video for Example 6 in the software and follow along in the space provided.

Example 6 Application: Calculating Living Expenses

Joe wants to budget $\frac{2}{5}$ of his monthly income for rent. He found an apartment he likes for $800 a month. What monthly income does he need to be able to afford this apartment?

Solution

✏ Now You Try It!

Use the space provided to work out the solution to the next example.

Example A Application: Calculating Living Expenses

Jim plans to budget $\frac{3}{7}$ of his monthly income to send his son Taylor to private school. If the school he'd like Taylor to attend costs $1200 a month, what monthly income does Jim need to be able to afford the school?

🔭 Looking Ahead

The following example involves finding three consecutive integers whose sum is a specified value. The key to solving this problem is setting it up correctly using an algebraic expression to denote each integer.

Example Preview

Find three consecutive integers whose sum is 579.

Solution

The first thing to do is decide how the three integers will be represented in the equation.

Let

$$n = \text{the first integer,}$$

$$n + 1 = \text{the second consecutive integer, and}$$

$$n + 2 = \text{the third consecutive integer.}$$

Write an equation that models the situation and solve.

$$n + (n + 1) + (n + 2) = 579$$
$$n + n + 1 + n + 2 = 579$$
$$3n + 3 = 579$$
$$3n = 576$$
$$n = \frac{576}{3}$$
$$n = 192 \quad \text{the first integer}$$
$$n + 1 = 193 \quad \text{the second integer}$$
$$n + 2 = 194 \quad \text{the third integer}$$

The three consecutive integers are 192, 193, and 194.

4.R.3 Exercises

Concept Check

True/False. Determine whether each statement is true or false. If a statement is false, explain how it can be changed so the statement will be true. (**Note:** There may be more than one acceptable change.)

1. If an odd integer is divided by 2, the remainder will be 1.

2. To find 3 consecutive odd integers, you could use n, $n + 1$, and $n + 3$.

3. Odd integers are integers that are divisible by 1.

4. Even integers are consecutive if each is 2 more than the previous even integer.

Practice

Read each problem carefully, translate the various phrases into algebraic expressions, set up an equation, and solve the equation.

5. Five less than a number is equal to 13 decreased by the number. Find the number.

6. Twice a number increased by 3 times the number is equal to 4 times the sum of the number and 3. Find the number.

7. Find three consecutive integers whose sum is 93.

8. Find three consecutive odd integers such that the sum of twice the first and three times the second is 7 more than twice the third.

Applications

Solve.

9. **Expenses:** A collect call from a landline in Ohio to another landline in Ohio has a connection fee of $2.75 and a charge of $0.36 per minute. Mr. Anderson made a collect call which cost the receiver of the call $9.95. This situation can be modeled by $9.95 = $2.75 + $0.36m.

   a. The unknown value is represented by the variable m in the equation. What is the unknown value in this situation?

   b. Solve the equation for the variable.

   c. What does the answer to Part **b.** mean? Write a complete sentence.

10. **Event Planning:** Robin is in charge of purchasing desserts for a dinner party that her nonprofit organization is throwing. She decides to buy a cake and several specialty cupcakes from Barbara's Bombtastic Bakery. She needs to buy one 8-inch round cake which costs $19.50. She has $45 to spend and will spend the leftover amount on cupcakes, which are $8.50 for a box of 4. How many boxes of cupcakes can Robin purchase?

    a. What is the unknown value in this problem? Let the variable c represent this unknown value.

    b. Write an equation to represent this situation.

    c. Solve the equation for the variable.

    d. What does the answer to Part **c.** mean? Write a complete sentence.

Writing & Thinking

11. **a.** How would you represent four consecutive odd integers?

    b. How would you represent four consecutive even integers?

    c. Are these representations the same? Explain.

4.R.4 Greatest Common Factor (GCF) and Factoring by Grouping

Making Connections

Maybe you want to determine the maximum height of a projectile launched in the air or the price of a product that is so high it makes a company's revenue equal to zero. For these types of applications, we are often required to solve a polynomial equation. The ability to factor a polynomial using the GCF and grouping can be applied when solving polynomial equations.

In this section, you will learn skills that you can apply when answering questions like these:

- Find the x-intercepts, if any, of the graph of the following function

$$a(x) = -2x^2 + 4x$$

- Solve the following polynomial equation by factoring.

$$x^3 - x^2 = 72x$$

Building Foundations

The **greatest common factor (GCF)** of two or more integers is the _____

_____

Procedure for Finding the GCF of a Set of Terms

1. Find the prime factorization of _____
2. List all the factors that are _____
3. Raise each common factor to the _____
4. Multiply these powers to _____

Note: If there is no common prime factor or variable, then _____

PROCEDURE

Factoring Out the GCF

1. Find the GCF of the _____
2. Divide this monomial factor into _____

The product of the GCF and this new polynomial factor is _____

PROCEDURE

4.R.4   Greatest Common Factor (GCF) and Factoring by Grouping

▶ Watch and Work

Watch the video for Example 10 in the software and follow along in the space provided.

Example 10   Factoring Polynomials by Grouping

Factor $xy + 5x + y + 5$ by grouping.

Solution

✏ Now You Try It!

Use the space provided to work out the solution to the next example.

Example A   Factoring Polynomials by Grouping

Factor $x^2 + xy + x + y$ by grouping.

4.R.4 Greatest Common Factor (GCF) and Factoring by Grouping

🔭 Looking Ahead

The skills you have reviewed in this section are the foundation to the process for finding *x*-intercepts of a quadratic function, as the following example shows.

Example Preview

Find the *x*-intercepts, if any, of the graph of the following function.

$$f(x) = -3x^2 - 6x$$

Solution

Since the *x*-intercepts are the points on the *x*-axis where $f(x) = 0$, we need to solve the following equation.

$$-3x^2 - 6x = 0$$

At this point, we need to recognize that both of the terms on the left side of this quadratic equation have a common factor of $-3x$. So, this quadratic equation can be solved in the following manner.

$$-3x^2 - 6x = 0$$
$$-3x(x + 2) = 0$$
$$-3x = 0, \quad x + 2 = 0$$
$$x = 0, \quad \quad x = -2$$

Therefore, the parabola, which is the graph represented by this quadratic function, crosses the *x*-axis at $(-2, 0)$ and $(0, 0)$.

4.R.4 Exercises

Concept Check

True/False. Determine whether each statement is true or false. If a statement is false, explain how it can be changed so the statement will be true. (**Note:** There may be more than one acceptable change.)

1. When finding the GCF of a polynomial, you need to consider only the coefficients.

2. An expression is factored completely if none of its factors can be factored.

4.R.4 Greatest Common Factor (GCF) and Factoring by Grouping

3. One way to find the GCF of a set of numbers is to use the prime factorization of each number.

4. Binomials cannot be factored out of algebraic expressions.

Practice

Find the GCF for each set of terms.

5. $\{25, 30, 75\}$

6. $\{8a^3, 16a^4, 20a^2\}$

Factor each polynomial by finding the GCF (or $-1 \cdot$ GCF).

7. $14x + 21$

8. $10x^2y - 25xy$

Factor each of the polynomials by grouping. If a polynomial cannot be factored, write "not factorable."

9. $3x + 3y + ax + ay$

10. $10xy - 2y^2 + 7yz - 35xz$

Applications

Solve.

11. **Projectile Motion:** A circus performer is shot vertically into the air with an initial velocity of 48 feet per second. The height of the performer above the ground in feet can be described by the polynomial $48x - 16x^2$ after x seconds.

    a. Find the height of the circus performer after 2 seconds.

    b. Factor the polynomial $48x - 16x^2$.

    c. Use the factored form of the polynomial from Part **b.** to find the height of the circus performer after 2 seconds.

    d. Are the answers from Parts **a.** and **c.** the same? Explain why or why not.

Writing & Thinking

12. Explain why the GCF of $-3x^2 + 3$ is 3 and not -3.

4.R.4 Greatest Common Factor (GCF) and Factoring by Grouping

4.R.5 Factoring Trinomials: $x^2 + bx + c$

↻ Making Connections

The ability to factor trinomials of the form $x^2 + bx + c$ is of fundamental importance to the study of quadratic functions and rational functions. It allows you to efficiently find the implied domain and zeros of such functions.

In this section, you will learn skills that you can apply when answering questions like these:

- Find the x-intercepts, if any, of the graph of the following function.

$$r(x) = \frac{x^2 + 8x + 16}{2}$$

- Determine the implied domain of the following function.

$$h(x) = \frac{5x}{x^2 - 3x - 10}$$

- The total cost of producing a type of boat is given by $C(x) = 16000 - 80x + 0.1x^2$, where x is the number of boats produced. How many boats should be produced to incur minimum cost?

✖ Building Foundations

> ### To Factor Trinomials of the Form $x^2 + bx + c$
>
> To factor $x^2 + bx + c$, if possible, find _____
>
> 1. If c is positive, then _____
>   a. Both will be _____
>   Example: _____
>   b. Both will be _____
>   Example: _____
> 2. If c is negative, _____
>   Examples: $x^2 + 6x - 7 = (x+7)(x-1)$ and _____
>
> PROCEDURE

4.R.5 Factoring Trinomials: $x^2 + bx + c$

▶ Watch and Work

Watch the video for Example 1 in the software and follow along in the space provided.

Example 1 **Factoring Trinomials with Leading Coefficients of 1**

Factor: $x^2 + 8x + 12$

Solution

✏ Now You Try It!

Use the space provided to work out the solution to the next example.

Example A **Factoring Trinomials with Leading Coefficients of 1**

Factor: $x^2 + 10x + 21$

🔭 Looking Ahead

Now that you have reviewed the factorization of trinomials of the form $x^2 + bx + c$, we will see how these ideas can be used in an application to so-called "break-even" analysis: we want to determine how much a company should charge for a good so that the cost of operating the business is the same as its revenue.

Example Preview

A restaurant frequently offers a special prix fixe meal and has been charging $130 per person for the event. At that price, they've been averaging 40 customers each time. Their marketing firm has convinced them that they'll gain a customer for every dollar they lower the cost of the event, and conversely lose a customer for every dollar they raise the cost. Their fixed cost per event is $2600 and preparing each customer's meal costs an additional $20. What are the break-even points in terms of customers served?

Solution

First, we will determine the cost and revenue functions. The cost function is the fixed cost of $2600 plus another $20 for every customer. If we let x represent the number of customers, then the cost function for holding an event for x persons is $C(x) = 2600 + 20x$.

We know that the restaurant gains a customer for every dollar they lower the cost of the event and loses a customer for every dollar they raise the cost. Thus, if they charge $130 + n$ dollars for the event, they can expect $40 - n$ customers on average. Since we let x denote the number of customers, we have $x = 40 - n$, or $n = 40 - x$. Then, the charge per person is $130 + 40 - x = 170 - x$ and the revenue function is $R(x) = x(170 - x)$.

The break even points are those values of x for which $C(x) = R(x)$. We can find these points by factoring.

$$2600 + 20x = x(170 - x)$$
$$2600 + 20x - 170x + x^2 = 0$$
$$x^2 - 150x + 2600 = 0$$
$$(x - 20)(x - 130) = 0$$

Thus, the break-even points are $x = 20$ and $x = 130$.

4.R.5 Exercises

Concept Check

True/False. Determine whether each statement is true or false. If a statement is false, explain how it can be changed so the statement will be true. (**Note:** There may be more than one acceptable change.)

1. In a trinomial such as $x^2 - 5x + 4$, one would need to find two factors of 4 whose sum is negative 5.

2. In factoring a trinomial with leading coefficient 1, if the constant term is negative, then both factors must be negative.

3. The first step in factoring a trinomial is to look for a common monomial factor.

4. For a trinomial with leading coefficient 1, if no pair exists whose product is the constant and whose sum is the middle term's coefficient, then the trinomial is not factorable.

Practice

Completely factor each trinomial. If a trinomial cannot be factored, write "not factorable."

5. $x^2 - 6x - 27$

6. $a^2 + a + 2$

7. $y^2 - 14y + 24$

8. $2a^4 + 24a^3 + 54a^2$

Applications

Solve.

9. **Triangles:** The area of a triangle is $\frac{1}{2}$ the product of its base and its height. If the area of the triangle shown is given by the function $A(x) = \frac{1}{2}x^2 + 24x$, find representations for the lengths of its base and its height (where the base is longer than the height).

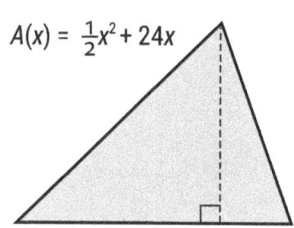

10. **Rectangles:** The area of the rectangle shown is given by the polynomial function $A(x) = 4x^2 + 20x$. If the width of the rectangle is $4x$, what is the length?

4.R.5 Factoring Trinomials: $x^2 + bx + c$

Writing & Thinking

11. Discuss, in your own words, how the sign of the constant term determines what signs will be used in the factors when factoring trinomials.

4.R.6 Factoring Trinomials: $ax^2 + bx + c$

↻ Making Connections

Factoring trinomials of the form $ax^2 + bx + c$ allows you to solve some particular types of quadratic equations without the need of more elaborate methods.

In this section, you will learn skills that you can apply when answering questions like these:

- Find the *x*-intercepts of the following function.

$$f(x) = 3x^2 + 5x - 2$$

- Determine the implied domain of the following function.

$$f(x) = \frac{x+1}{2x^2 - 3x + 1}$$

- The height of a rock t seconds after breaking free from the top of a 260-foot cliff is given by $h(t) = -16t^2 + 260$. Determine the time it takes the rock to hit the bottom of the cliff.

🛠 Building Foundations

> Guidelines for the Trial-and-Error Method
> 1. If the sign of the constant term is positive (+), the signs in _____
>
> 2. If the sign of the constant term is negative (–), the signs in _____

4.R.6 Factoring Trinomials: $ax^2 + bx + c$

Analysis of Factoring by the ac-Method

	General Method	**Example**
	$ax^2 + bx + c$	$2x^2 + 9x + 10$
Step 1:	Multiply _____	Multiply _____
Step 2:	Find two integers whose product is ac and _____ _____ _____	Find two integers whose product is 20 and _____ _____ _____
Step 3:	Rewrite the middle term (bx) using the _____ _____ _____	_____ _____ _____
Step 4:	Factor by grouping the _____ _____	Factor by grouping the _____ _____
Step 5:	Factor out the common binomial factor. This will give _____ _____	Factor out the common binomial factor _____ _____

PROCEDURE

▶ Watch and Work

Watch the video for Example 3 in the software and follow along in the space provided.

Example 3 Using the *ac*-Method

Use the *ac*-method to factor $3x^2 + 19x + 6$.

Solution

4.R.6 Factoring Trinomials: $ax^2 + bx + c$

✏️ Now You Try It!

Use the space provided to work out the solution to the next example.

Example A Using the *ac*-Method

Use the *ac*-method to factor $3a^2 + 14a + 8$.

🔭 Looking Ahead

Now that you have reviewed the factorization of trinomials of the form $ax^2 + bx + c$, you will see how these ideas can be used in a physics application involving free-fall, as shown in the following example.

Example Preview

The height of a rock t seconds after breaking free from the top of a 196-foot cliff is given by $h(t) = -16t^2 + 196$. Determine the time it takes the rock to hit the bottom of the cliff.

Solution

The height of the rock is zero at the bottom of the cliff, $h(t) = 0$. Therefore, we can rewrite the function and solve for t.

$$h(t) = -16t^2 + 196$$
$$0 = -16t^2 + 196$$
$$0 = -4(4t^2 - 49)$$
$$0 = 4t^2 - 49$$
$$0 = (2t - 7)(2t + 7)$$

We then set each factor equal to zero and solve for t.

$$2t - 7 = 0 \qquad 2t + 7 = 0$$
$$2t = 7 \qquad 2t = -7$$
$$t = \frac{7}{2} = 3.5 \qquad t = \frac{-7}{2} = -3.5$$

Since time in this instance cannot be negative, the correct answer is $t = 3.5$ seconds. It took the rock 3.5 seconds to fall 196 feet to the bottom of the cliff.

4.R.6 Exercises

Concept Check

True/False. Determine whether each statement is true or false. If a statement is false, explain how it can be changed so the statement will be true. (**Note:** There may be more than one acceptable change.)

1. A trinomial is factorable if the middle term is the difference of the inner and outer products of two binomials.

2. The trial-and-error method of factoring a trinomial follows the same steps as the FOIL method of multiplication.

3. The first step in the *ac*-method of factoring is to rewrite the middle term.

4. Factoring can be checked by multiplying the factors and verifying that the product matches the original polynomial.

Practice

Completely factor each polynomial. If a polynomial cannot be factored, write "not factorable."

5. $6x^2 + 11x + 5$

6. $-x^2 + 3x - 2$

4.R.6 Factoring Trinomials: $ax^2 + bx + c$

7. $x^2 + 8x + 64$

8. $9x^2 - 3x - 20$

9. $12x^2 - 38x + 20$

10. $5a^2 - 7a + 2$

Writing & Thinking

11. It is true that $2x^2 + 10x + 12 = (2x+6)(x+2) = (2x+4)(x+3)$. Explain how the trinomial can be factored in two ways. Is there some kind of error?

12. It is true that $5x^2 - 5x - 30 = (5x-15)(x+2)$. Explain why this is not the completely factored form of the trinomial.

4.R.7 Review of Factoring Techniques

⟳ Making Connections

Being able to determine which factoring method to use when attempting to solve an equation is a skill that is best developed with practice over time. It is also important to be able to tell when an equation cannot be solved via factoring and should be solved via other means. Therefore, learning all the factoring methods and practicing them will make it significantly easier for you adequately determine methodology for solving polynomial equations.

In this section, you will learn skills that you can apply when answering questions like these:

Solve the following polynomial equations.
- $x^4 - 29x^2 + 100 = 0$
- $x^4 + 7x^2 = 8$

🛠 Building Foundations

General Guidelines for Factoring Polynomials

1. **Always look for a common** _____

    If the leading coefficient is negative, _____

2. **Check the number of terms.**

    a. **Two terms:**

    1. difference of _____
    2. sum of _____
    3. difference of _____
    4. sum of _____

    b. **Three terms:**

    1. _____
    2. _____

    Guidelines for the trial-and-error method

    a. If the sign of the constant term is positive, the signs in _____
    _____

    b. If the sign of the constant term is negative, the signs in _____
    _____

(continues...)

PROCEDURE

General Guidelines for Factoring Polynomials (...continued)

3. _____

   Guidelines for the *ac*-method

   a. Multiply _____

   b. Find two integers whose product is *ac* and _____

      _____

   c. Rewrite the middle term (*bx*) using _____

   d. _____

   c. **Four terms:**

      Group terms with a common factor and _____

3. Check the possibility of _____

   Checking: Factoring can be checked by _____

   The product should be _____

PROCEDURE

🔭 Looking Ahead

In the following example, you will use the factoring skills that you have reviewed so far to solve a fourth degree polynomial equation.

Example Preview

Solve the following polynomial equation by factoring or using the quadratic formula. Identify all solutions.

$$x^4 - 13x^2 + 36 = 0$$

Solution

Noticing that x^4 can be factored into x^2 times x^2 and that $36 = (-9)(-4)$ and $(-9)+(-4) = -13$, the polynomial can be factored into two quadratics.

$$x^4 - 13x^2 + 36 = (x^2 - 9)(x^2 - 4) = 0$$

The two quadratics are differences of perfect squares which can be further factored as follows.

$$x^4 - 13x^2 + 36 = (x^2 - 9)(x^2 - 4) = (x+2)(x-2)(x+3)(x-3) = 0$$

Setting the four linear factors equal to zero and solving for x yields:

$$x + 2 = 0 \qquad x - 2 = 0 \qquad x + 3 = 0 \qquad x - 3 = 0$$
$$x = -2 \qquad x = 2 \qquad x = -3 \qquad x = 3$$

This gives us $\{\pm 2, \pm 3\}$ as the solution set of the equation.

4.R.7 Exercises

Concept Check

True/False. Determine whether each statement is true or false. If a statement is false, explain how it can be changed so the statement will be true. (**Note:** There may be more than one acceptable change.)

1. You should always start by checking the number of terms when factoring a polynomial.

2. If a trinomial is to be factored, the trial-and-error or *ac*-methods can be used.

3. If there are four terms in a polynomial, it cannot be factored.

Practice

Completely factor each of the given polynomials. If a polynomial cannot be factored, write "not factorable."

4. $x^2 - 100$

5. $x^2 + 10x + 25$

6. $x^2 + 16x + 64$

7. $20x^2 - 21x - 54$

8. $2y^2 + 6yz + 5y + 15z$

9. $x^3 + 125$

10. $a^2 + 2a + 24$

11. $x^2 + 9x - 36$

12. $64 + 49t^2$

13. $4x^2 - 14x + 6$

14. $200x + 20x^2 - 4x^3$

4.R.8 Solving Quadratic Equations by Factoring

⟳ Making Connections

Solving a quadratic equation often involves factoring a quadratic polynomial using one of the many techniques you have reviewed so far.

In this section, you will learn skills that you can apply when answering questions like these:

- Find the *x*-intercepts, if any, of the graphs of the following functions.

$$q(x) = x^2 + 8x + 15$$

$$g(x) = -2x^2 - 8x$$

✄ Building Foundations

Quadratic Equations

Quadratic equations are equations that can be written in the form

_____ where a, b, and c are real numbers and $a \neq 0$.

DEFINITION

Zero-Factor Property

If the product of two (or more) factors is 0, then _____

That is, for real numbers a and b,

if $a \cdot b = 0$, then a _____

DEFINITION

4.R.8 Solving Quadratic Equations by Factoring

▶ Watch and Work

Watch the video for Example 4 in the software and follow along in the space provided.

Example 4 Solving Quadratic Equations by Factoring

Solve by factoring: $4x^2 - 4x = 24$

Solution

✏ Now You Try It!

Use the space provided to work out the solution to the next example.

Example A Solving Quadratic Equations by Factoring

Solve by factoring: $9x^2 - 27x = 36$

4.R.8 Solving Quadratic Equations by Factoring

To Solve a Quadratic Equation by Factoring

1. Add or subtract terms as necessary so that _____ and the equation is in the _____ where a, b, and c are real numbers and $a \neq 0$.

2. Factor completely. (If there are any fractional coefficients, _____ _____

3. Set each nonconstant factor equal to _____

4. Check each solution, one at a time, in _____

PROCEDURE

Factor Theorem

If $x = c$ is a root of a polynomial equation in the form _____
_____

THEOREM

🔭 Looking Ahead

Now that you have reviewed solving quadratic equations by factoring, you will see how these ideas can be used when solving polynomials equations of degree higher than two. The follow example reduces solving a polynomial equation of degree three by factoring.

Example Preview

Solve the following polynomial equation.

$$x^3 - 74x^2 + 73x = 0$$

Solution

$$x^3 - 74x^2 + 73x = 0$$
$$x(x^2 - 74x + 73) = 0$$
$$x(x - 73)(x - 1) = 0$$
$$x = 0, 73, 1$$

4.R.8 Exercises

Concept Check

True/False. Determine whether each statement is true or false. If a statement is false, explain how it can be changed so the statement will be true. (**Note:** There may be more than one acceptable change.)

1. When solving quadratic equations by factoring, it is important that all of the coefficients are integers.

2. The standard form for a quadratic equation is $ax^2 + bx = c$.

3. Not all quadratic equations can be solved by factoring.

4. All quadratic equations have two distinct solutions.

Practice

Solve each equation by factoring.

5. $x^2 - 11x + 18 = 0$

6. $9x^2 + 63x + 90 = 0$

7. $(x-5)(x+3) = 9$

8. Find a polynomial equation with integer coefficients that has $x = 5$ and $x = 7$ as roots.

Applications

Solve.

9. **Falling Objects:** A ball is dropped from the top of a building that is 784 feet high. The height of the ball above ground level is given by the polynomial function $h(t) = -16t^2 + 784$ where t is measured in seconds.

   a. How high is the ball after 3 seconds? 5 seconds?

   b. How far has the ball traveled in 3 seconds? 5 seconds?

   c. When will the ball hit the ground? Explain your reasoning in terms of factors.

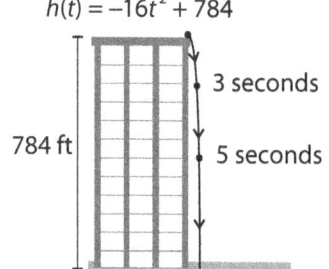

10. **Falling Objects:** A tennis ball is dropped from a building. The position of the ball after t seconds is given by the polynomial function $s(t) = -4.9t^2 + 490$, where s is the height in meters of the ball.

    a. Find $s(0)$. What does this value represent in the context of this problem?

    b. How high is the tennis ball 2 seconds after it has been dropped?

    c. How long before the tennis ball hits the ground?

Writing & Thinking

11. When solving equations by factoring, one side of the equation must be 0. Explain why this is so.

12. In solving the equation $(x+5)(x-4) = 6$, why can't we just put one factor equal to 3 and the other equal to 2? Certainly $3 \cdot 2 = 6$.

4.R.8 Solving Quadratic Equations by Factoring

4.R.9 Multiplication and Division with Complex Numbers

♻ Making Connections

Multiplying and dividing complex numbers is a necessary skill to tackle many problems involving the square root of a negative number. Complex numbers allow you to work in a theoretical field of math and find solutions otherwise missed.

In this section, you will learn skills that you can apply when answering questions like these:

- Simplify the following expression.

$$i^{26}$$

- Evaluate the following square root expression.

$$(-i)^6 \sqrt{-225}$$

- Given that $2 - 3i$ is a zero, factor the following polynomial function completely.

$$f(x) = x^4 - 7x^3 + 27x^2 - 47x + 26$$

🛠 Building Foundations

The two complex numbers $a + bi$ and $a - bi$ are called _____ or simply **conjugates** of each other. **the product of two complex conjugates will always be a** _____

Writing Fractions with Complex Numbers in Standard Form

1. Multiply both the numerator and denominator _____
2. Simplify the resulting _____
3. Write the _____

PROCEDURE

4.R.9 Multiplication and Division with Complex Numbers

▶ Watch and Work

Watch the video for Example 1 in the software and follow along in the space provided.

Example 1 Multiplying with Complex Numbers

Find the following products.

a. $(3i)(2-7i)$

b. $(5+i)(2+6i)$

c. $(\sqrt{2}-i)(\sqrt{2}-i)$

d. $(-1+i)(2-i)$

Solution

✏️ Now You Try It!

Use the space provided to work out the solution to the next example.

Example A Multiplying with Complex Numbers

Find the following products.

a. $(4i)(1-7i)$

b. $(4+2i)(3+5i)$

c. $(\sqrt{2}-i)(\sqrt{2}+i)$

🔭 Looking Ahead

Now that you have reviewed multiplication and division with complex numbers, you have the tools necessary to work with polynomial functions that have complex zeros. Consider the following example.

Example Preview

Construct a polynomial function f with the following properties: third degree, only real coefficients, -2 and $2+i$ are two of the zeros, y-intercept is -10.

Solution

As $2+i$ is one of the zeros and the polynomial function is to have only real coefficients, we know by the Conjugate Roots Theorem that $2-i$ must be a zero as well. Based on this, f must be of the form

$$f(x) = a(x-(2+i))(x-(2-i))(x+2)$$

for some real constant a.

Because the *y*-intercept is -10, the coefficient *a* must be chosen so that $f(0) = -10$. To make this calculation simpler, begin my multiplying out $(x-(2+i))(x-(2-i))$.

$$(x-(2+i))(x-(2-i)) = (x-2-i)(x-2+i)$$
$$= x^2 - 2x + ix - 2x + 4 - 2i - ix + 2i - i^2$$
$$= x^2 - 2x + \cancel{ix} - 2x + 4 \cancel{- 2i} \cancel{- ix} \cancel{+ 2i} - (-1)$$
$$= x^2 - 2x - 2x + 4 + 1$$
$$= x^2 - 4x + 5$$

Now, substitute $f(0) = -10$ and solve for *a*.

$$f(x) = a(x^2 - 4x + 5)(x+2)$$
$$f(0) = a((0)^2 - 4(0) + 5)((0) + 2)$$
$$-10 = a(5)(2)$$
$$-10 = 10a$$
$$a = -1$$

Thus, the simplified polynomial function is as follows.

$$f(x) = (-1)(x^2 - 4x + 5)(x+2)$$
$$= -(x^3 + 2x^2 - 4x^2 - 8x + 5x + 10)$$
$$= -x^3 + 2x^2 + 3x - 10$$

4.R.9 Exercises

Concept Check

True/False. Determine whether each statement is true or false. If a statement is false, explain how it can be changed so the statement will be true. (**Note:** There may be more than one acceptable change.)

1. Regardless of the value of the exponent, the only possible values for any power of *i* are *i* and $-i$.

2. The product $\sqrt{a} \cdot \sqrt{b}$ can be rewritten as $\sqrt{ab}$ as long as *a* and *b* are real numbers.

3. When *i* is squared, the product is 1.

4. The conjugate of 4 − 5*i* is 4 + 5*i*.

Practice
Perform the indicated operations and write each result in standard form.

5. $-4i(6-7i)$

6. $(2+7i)(6+i)$

7. $\dfrac{5}{4i}$

8. $\dfrac{6+i}{3-4i}$

4.R.9 Multiplication and Division with Complex Numbers

Simplify the following powers of *i* and write each result in standard form. Assume *k* is a positive integer.

9. i^{13}

10. i^{-3}

Find the indicated products and simplify.

Writing & Thinking

11. Explain why the product of every complex number and its conjugate is a nonnegative real number.

12. What condition is necessary for the conjugate of a complex number, $a + bi$, to be equal to the reciprocal of this number?

4.R.10 Quadratic Equations: The Quadratic Formula

↻ Making Connections

While techniques used to factor a trinomial of the form $ax^2 + bx + c$ can be very useful when solving quadratic equations, sometimes the equations become too cumbersome and we need a more general technique. The quadratic formula is the most common of such techniques.

In this section, you will learn skills that you can apply when answering questions like these:

- Solve the following quadratic equation using the quadratic formula.

$$-3x^2 + 7x = 1$$

- Determine the values of the variable that must be excluded for the following rational expression.

$$\frac{5x^2 - 24x - 36}{x^2 - 4x - 12}$$

- Solve the following radical equation.

$$\sqrt{3x+7} + 8 = x + 6$$

🛠 Building Foundations

> **The Quadratic Formula**
> For the general quadratic equation _____
>
> _____
>
> FORMULA

1. The expression $b^2 - 4ac$, the part of the quadratic formula that lies under the radical sign, is called the _____.

Discriminant	Nature of Solutions
$b^2 - 4ac > 0$	
$b^2 - 4ac = 0$	
$b^2 - 4ac < 0$	

Table 1

▶ Watch and Work

Watch the video for Example 2 in the software and follow along in the space provided.

Example 2 The Quadratic Formula

Solve by using the quadratic formula.

$7x^2 - 2x + 1 = 0$

Solution

✏ Now You Try It!

Use the space provided to work out the solution to the next example.

Example A The Quadratic Formula

Solve by using the quadratic formula.

$5x^2 - 3x + 2 = 0$

Looking Ahead

The ability to find the zeros of a quadratic equation using the quadratic formula are applied especially when the quadratic equation is not easily factored, as shown in the following example.

Example Preview

Solve the following polynomial equation.

$$9x^2 - 49x + 39 = 0$$

Solution

The quadratic formula is

$$x = \frac{-b \pm \sqrt{b^2 - 4ac}}{2a}$$

where the general quadratic equation is of the form $ax^2 + bx + c = 0$ and $a \neq 0$.

$$9x^2 - 49x + 39 = 0$$

$$x = \frac{-(-49) \pm \sqrt{(49)^2 - 4(9)(39)}}{2(9)}$$

$$x = \frac{49 \pm \sqrt{2401 - 1404}}{18}$$

$$x = \frac{49 \pm \sqrt{997}}{18}$$

This gives us $\dfrac{49 + \sqrt{997}}{18}$ and $\dfrac{49 - \sqrt{997}}{18}$ as the two solutions of the equation.

4.R.10 Exercises

Concept Check

True/False. Determine whether each statement is true or false. If a statement is false, explain how it can be changed so the statement will be true. (**Note:** There may be more than one acceptable change.)

1. The quadratic formula will always work when solving quadratic equations.

2. If the discriminant is a perfect square, the quadratic equation is factorable.

3. When using the quadratic formula, if the discriminant is greater than zero, there are infinite solutions.

4. If the discriminant is less than zero, there is no real solution.

Practice

Find the discriminant and determine the nature of the solutions of each quadratic equation.

5. $x^2 + 6x - 8 = 0$

6. $x^2 - 8x + 16 = 0$

Solve each of the quadratic equations using the quadratic formula.

7. $x^2 + 4x - 4 = 0$

8. $x^2 - 2x + 7 = 0$

9. $3x^2 - 7x + 4 = 0$

Applications

Solve.

10. **Throwing Objects:** An orange is thrown down from the top of a building that is 300 feet tall with an initial velocity of 6 feet per second. The distance of the object from the ground can be calculated using the equation $d = 300 - 6t - 16t^2$, where t is the time in seconds after the orange is thrown.

   a. On a balcony, a cup is sitting on a table located 100 feet from the ground. If the orange is thrown with the right aim to fall into the cup, how long will the orange fall? Round to the nearest hundredth. (**Hint:** The distance is 100 feet.)

   b. If the orange misses the cup and falls to the ground, how long will it take for the orange to splatter on the sidewalk? (**Hint:** What is the height of the orange when it hits the ground?)

   c. Approximately how much longer would it take for the orange to fall to the sidewalk than it would for the orange to fall into the cup?

Writing & Thinking

11. Find an equation of the form $Ax^4 + Bx^2 + C = 0$ that has the four roots ±2 and ±3. Explain how you arrived at this equation.

12. The surface area of a right circular cylinder can be found using the following formula: $S = 2\pi r^2 + 2\pi rh$, where r is the radius of the cylinder and h is the height. Estimate the radius of a circular cylinder of height 30 cm and surface area 300 cm². Explain how you used your knowledge of quadratic equations.

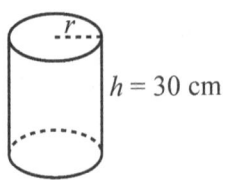

CHAPTER 5.R

Review Concepts

for *Working with Functions*

5.R.1 Order of Operations with Real Numbers

5.R.2 Simplifying and Evaluating Algebraic Expressions

5.R.3 Multiplication with Polynomials

5.R.4 Division with Polynomials

5.R.5 Introduction to Rational Expressions

5.R.6 Multiplication and Division with Rational Expressions

5.R.7 Simplifying Complex Fractions

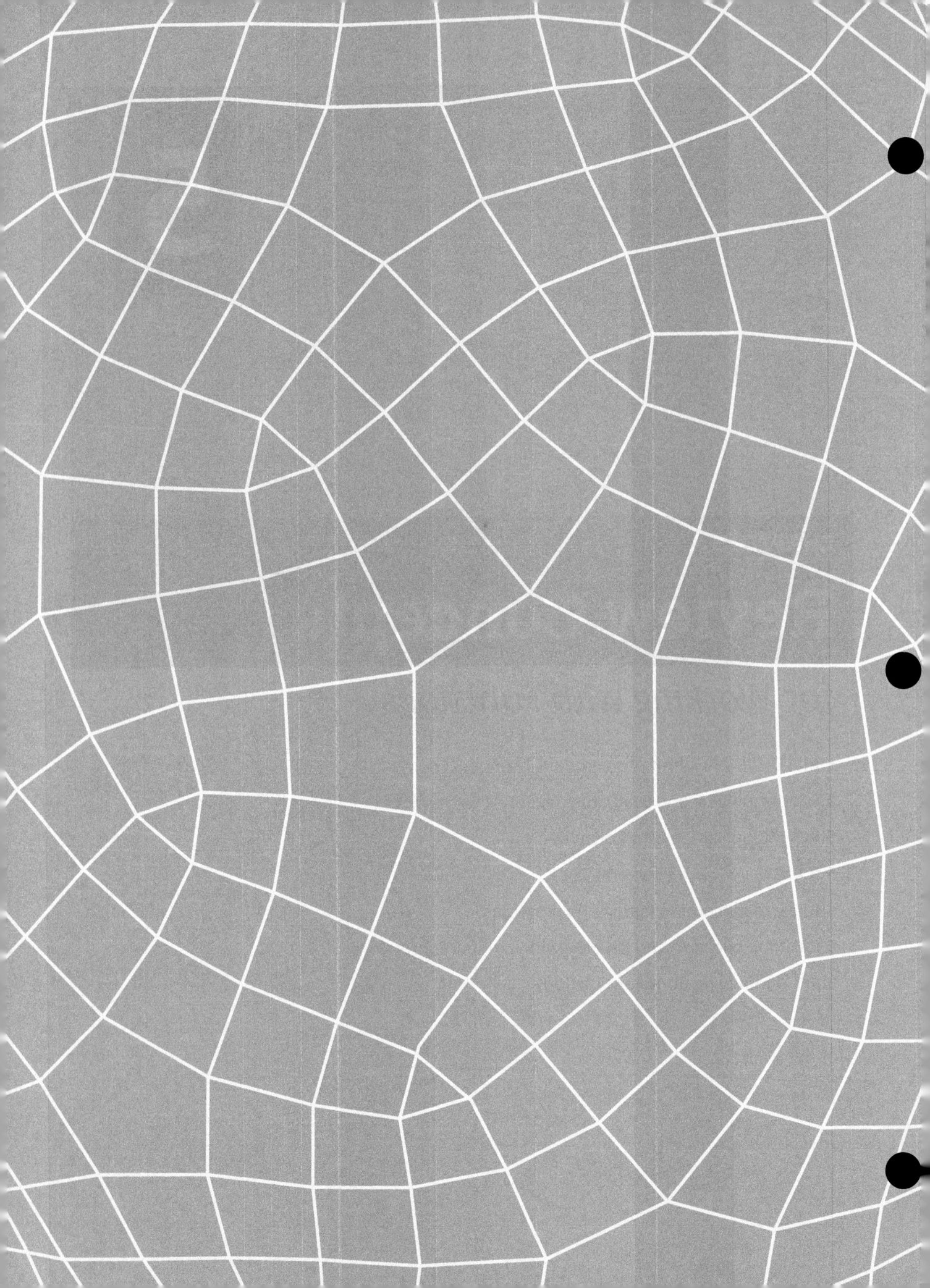

5.R.1 Order of Operations with Real Numbers

↻ Making Connections

To evaluate a function at a particular value, you must substitute that value in for the variable and then simplify the expression. To correctly simplify the expression, you must properly use the order of operations. Reviewing the order of operations will help to ensure you correctly evaluate functions and simplify expressions.

In this section, you will learn skills that you can apply when answering questions like these:

For each function below, determine **a.** $f(2)$, **b.** $f(x-1)$, **c.** $f(x+a) - f(x)$, and **d.** $f(x^2)$.

- $f(x) = x^2 + 3x$
- $f(x) = -x^2 - 7$

🛠 Building Foundations

> **Rules for Order of Operations**
> 1. Simplify within grouping symbols, such as _____
>   _____
> 2. Find any _____
> 3. Moving from left to right, perform any _____
> 4. Moving from left to right, perform any _____
>
> PROCEDURE

▶ Watch and Work

Watch the video for Example 3 in the software and follow along in the space provided.

Example 3 Using the Order of Operations with Real Numbers

Simplify: $(2-5)^2 + |2 - 5^2| - 2^3$

Solution

Now You Try It!

Use the space provided to work out the solution to the next example.

Example A Using the Order of Operations with Real Numbers

Simplify: $(1-3)^2 + |9 - 4^2| - 1^3$

Looking Ahead

Your review of the order of operations will be helpful in evaluating functions. Evaluating the function $g(x)$ in the following example involves exponents, multiplication, and subtraction, which must be done in the proper order.

Example Preview

Determine $g(x+a) - g(x)$ for the following function.

$$g(x) = 5x^2 - 4x$$

Solution

To determine $g(x+a) - g(x)$, we substitute the values in for every occurrence of x.

$$g(x+a) - g(x) = \left(5(x+a)^2 - 4(x+a)\right) - \left(5x^2 - 4x\right)$$
$$= 5\left(x^2 + 2ax + a^2\right) - 4(x+a) - \left(5x^2 - 4x\right) \quad \text{Apply the exponent.}$$
$$= 5x^2 + 10ax + 5a^2 - 4x - 4a - 5x^2 + 4x \quad \text{Multiply across the parentheses.}$$
$$= 10ax + 5a^2 - 4a \quad \text{Add and subtract.}$$

5.R.1 Exercises

Concept Check

True/False. Determine whether each statement is true or false. If a statement is false, explain how it can be changed so the statement will be true. (**Note:** There may be more than one acceptable change.)

1. If there are no grouping symbols, multiplication should always be performed before addition.

2. When following the rules for order of operations, powers indicated by exponents should be evaluated last.

3. The square root symbol is a grouping symbol.

4. A well-known mnemonic device for remembering the rules for order of operations is SADMEP.

Practice

Simplify.

5. a. $24 \div 4 \cdot 6$

   b. $24 \cdot 4 \div 6$

6. $15 \div (-3) \cdot 3 - 10$

7. $3^2 \div (-9) \cdot (4 - 2^2) + 5(-2)$

8. $14 \cdot 3 \div (-2) - 6(4)$

9. $|16 - 20| + (-10)^2 + 5^2$

10. $6(13 - 15)^2 \cdot 8 \div 2^2 + 3(-1)$

11. $8 \quad 9\left[(-39) \div (-13) + 7(-2) - (-2)^2\right]$

Applications

Solve.

12. ***Discounts:*** The Matthews family, a family of 4, is planning a trip to New York City. During their visit, they want to see the Broadway play *Matilda*. The tickets cost $102 each. The Matthews purchase the tickets online and the website charges a service fee of $7.50 per ticket. The website is running a sale where the Matthews can get 10% off of their entire purchase.

    a. Write an expression to describe how much of a discount the Matthews will receive on their purchase.

    b. What is the final purchase price of the tickets?

13. ***Banking:*** Dennis overdrew his checking account and ended up with a balance of −$42. The bank charged a $35 overdraft fee and an additional $5 fee for every day the account was overdrawn. Dennis left his account overdrawn for 3 days.

    a. Write an expression to show the balance of Dennis's checking account after 3 days.

    b. Simplify the expression in Part **a.** to find the balance of Dennis's checking account after 3 days.

Writing & Thinking

14. Explain, in your own words, why the following expression cannot be evaluated.

$$(24-2^4)+6(3-5) \div (3^2-9)$$

15. Consider any number between 0 and 1. If you square this number, will the result be larger or smaller than the original number? Is this always the case? Explain.

5.R.1 Order of Operations with Real Numbers

5.R.2 Simplifying and Evaluating Algebraic Expressions

↻ Making Connections

A function is usually easier to work with if it is written in simplified form. Actions such as evaluating, graphing, combining, and finding the inverse are easier if performed on simplified functions. Therefore, it is recommended that you review the process of simplifying expressions before proceeding.

In this section, you will learn skills that you can apply when answering questions like these:

- Given the two functions $g(x) = x^2 - 1$ and $f(x) = x - 1$, find $(f+g)(x)$ and $(fg)(x)$.
- Use the following function information to determine **a.** $(f+g)(-1)$, **b.** $(f-g)(-1)$, **c.** $(fg)(-1)$, and **d.** $\left(\dfrac{f}{g}\right)(-1)$.

$$f(x) = x^4 + 1 \text{ and } g(x) = x^{11} + 2$$

🛠 Building Foundations

> **Like Terms**
> Like terms (or similar terms) are terms that are _____
> _____
>
> **DEFINITION**

> **Combining Like Terms**
> To combine like terms, _____
>
> **DEFINITION**

> **To Evaluate an Algebraic Expression**
> 1. _____
> 2. _____
> 3. _____
>
> (**Note:** Terms separated by + and − signs may be evaluated _____
> _____.)
>
> **PROCEDURE**

▶ Watch and Work

Watch the video for Example 5 in the software and follow along in the space provided.

Example 5 Simplifying and Evaluating Algebraic Expressions

Simplify and evaluate $3ab - 4ab + 6a - a$ for $a = 2$, $b = -1$.

Solution

✏ Now You Try It!

Use the space provided to work out the solution to the next example.

Example A Identifying Properties of Addition and Multiplication

Simplify and evaluate

$5ab - 8ab + 2a - 3a$ for $a = -3$, $b = 1$

Looking Ahead

The skills you learned related to simplifying expressions will be very helpful when combining functions through addition, subtraction, multiplication, division, and composition. The following example shows how a combination of two functions is easier to understand once simplified.

Example Preview

Consider the following functions.

$$f(x) = x^3 + 3 \text{ and } g(x) = 4x$$

Find the formula for $(f+g)(x)$ and simplify your answer. Then find the domain for $(f+g)(x)$.

Solution

From the definitions of addition, subtraction, multiplication, and division of functions, we know that

$$\begin{aligned}(f+g)(x) &= f(x) + g(x) \\ &= (x^3 + 3) + (4x) \\ &= x^3 + 4x + 3.\end{aligned}$$

The domain of $(f+g)(x)$ is the entire set of real numbers because both $f(x)$ and $g(x)$ are defined for all real numbers.

5.R.2 Exercises

True/False. Determine whether each statement is true or false. If a statement is false, explain how it can be changed so the statement will be true. (**Note:** There may be more than one acceptable change.)

1. A variable that does not appear to have an exponent has an exponent of 1.

2. In the term $-9x$, nine is being subtracted from x.

3. In the term "$12a$," 12 is the constant.

4. Like terms have the same coefficients.

Practice

Identify the like terms in each list of terms.

5. $-5, 3, 7x, 8, 9x, 3y$

Simplify each expression by combining like terms.

6. $8x + 7x$

7. $3x - 5x + 12x$

8. $13x + 12x^2 + 15x - 35 - 41 - 2x^2$

Simplify each expression and then evaluate the expression for $y = 3$ and $a = -2$.

9. $5y + 4 - 2y$

10. $\dfrac{3a + 5a}{-2} + 12a$

Applications

Solve.

11. **Profit:** An apartment management company owns a property with 100 units. The company has determined that the profit made per month from the property can be calculated using the equation $P = -10x^2 + 1500x - 6000$, where x is the number of units rented per month. How much profit does the company make when 80 units are rented?

12. **Physics:** A ball is thrown upward from an initial height of 96 feet with an initial velocity of 16 feet per second. After t seconds, the height of the ball can be described by the expression $-16t^2 + 16t + 96$. What is the height of the ball after 3 seconds?

Writing & Thinking

13. Discuss like and unlike terms and give an example of each.

14. Explain the difference between -13^2 and $(-13)^2$.

5.R.3 Multiplication with Polynomials

🔄 Making Connections

Being able to multiply polynomials is an important skill that allows you to create new functions from old ones.

In this section, you will learn skills that you can apply when answering questions like these:

- Find the formula for $(fg)(x)$ for the following functions.

$$f(x) = x^3 + 1 \quad \text{and} \quad g(x) = x^2 - 3$$

- Given the following graph of two functions, find $(fg)(3)$.

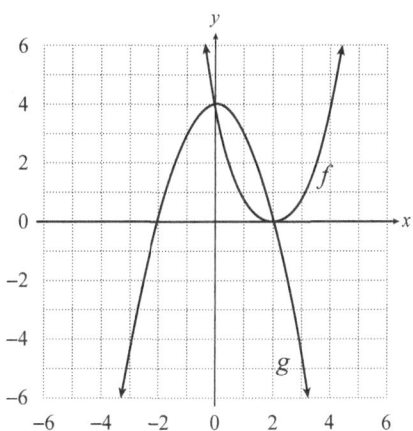

🛠 Building Foundations

Using the _____ property $a(b+c) = ab + ac$ with multiplication indicated on the left, we can find the product of a monomial with a polynomial of two or more terms as follows.

$$5x(2x+3) = \underline{} \cdot 2x + \underline{} \cdot 3 = \underline{}$$

We can apply the distributive property in the following way to multiply two polynomials.

$$(2x-1)(x^2+x-5) = \underline{}(x^2+x-5)\underline{}(x^2+x-5)$$
$$= 2x \cdot \underline{} + 2x \cdot \underline{} + 2x \cdot (\underline{}) - 1 \cdot \underline{} - 1 \cdot \underline{} - 1(\underline{})$$
$$= \underline{}$$
$$= \underline{}$$

5.R.3 Multiplication with Polynomials

▶ Watch and Work

Watch the video for Example 8 in the software and follow along in the space provided.

Example 8 Multiplying Polynomials

Multiply: $(x-5)(x+2)(x-1)$

Solution

✏ Now You Try It!

Use the space provided to work out the solution to the next example.

Example A Multiplying Polynomials

Multiply: $(x-2)(x+3)(x-6)$

Looking Ahead

Now that you have reviewed how to multiply polynomials, you will see how this idea can be used to perform operations with functions.

Example Preview

Find the formula for $(fg)(x)$ for the following functions.

$$f(x) = x^2 + 5 \quad \text{and} \quad g(x) = x^3 - 2$$

Solution

$$\begin{aligned}(fg)(x) &= f(x) \cdot g(x) \\ &= (x^2 + 5)(x^3 - 2) \\ &= x^6 - 2x^2 + 5x^3 - 10 \\ &= x^6 + 5x^3 - 2x^2 - 10\end{aligned}$$

5.R.3 Exercises

Concept Check

True/False. Determine whether each statement is true or false. If a statement is false, explain how it can be changed so the statement will be true. (**Note:** There may be more than one acceptable change.)

1. The distributive property can only be used to multiply a monomial and a polynomial.

2. The product of $(a + b)$ and $(c + d)$ is $ac + bd$.

3. The FOIL method is a way to remember one specific order that the distributive property can be applied.

Practice
Multiply and simplify, if necessary.

4. $-3x^2(2x^3+5x)$

5. $-4x^3(x^5-2x^4+3x)$

6. $(x+4)(x-3)$

7. $(y+3)(y^2-y+4)$

Applications

Solve.

8. *Advertising:* A graphic artist is designing a poster to advertise an upcoming event. The only restrictions regarding the poster size is that it must have a length of $3x$ inches and a width of $2x + 5$ inches. Find a simplified expression for the area of the poster.

9. *Shipping:* Armon works for a company that ships artwork worldwide. The size of each item varies, but all of the art is on square canvases. Armon's job is to make the wooden shipping crates for each piece of art. In order to protect the artwork, each crate must be 10 inches deep. The crate must also be 10 inches wider and 12 inches taller than the artwork. Letting x represent the length of one side of the artwork, find the volume of the rectangular shipping crate.

Writing & Thinking

10. We have seen how the distributive property is used to multiply polynomials.

    Show how the distributive property can be used to find the product

    $$\begin{array}{r} 75 \\ \times\, 93 \\ \hline \end{array}$$

    (**Hint:** $75 = 70 + 5$ and $93 = 90 + 3$)

5.R.4 Division with Polynomials

↻ Making Connections

The standard algorithm for dividing polynomials is very similar to the long division of integers, in that it allows us to produce a quotient of degree lower than that of the dividend. This is very important when trying to understand the behavior of rational functions.

In this section, you will learn skills that you can apply when answering questions like these:

- Find the equation(s) of the vertical asymptotes of the following rational function.

$$g(x) = \frac{x^2 - x - 2}{x + 1}$$

- Find the equation(s) of the horizontal or oblique asymptotes of the following rational function.

$$f(x) = \frac{-x^2 + 41x + 56}{-6x - 7}$$

✖ Building Foundations

To divide the numerator by the denominator (with a monomial in the denominator), we _____ _____ and simplify each fraction.

The Division Algorithm

For polynomials P and D, the division algorithm gives

$$\underline{\hspace{3cm}}$$

where Q and R are polynomials and the degree of _____

THEOREM

▶ Watch and Work

Watch the video for Example 4 in the software and follow along in the space provided.

Example 4 Using Long Division (Terms Missing)

Simplify $\dfrac{x^4 + 9x^2 - 3x + 5}{x^2 - x + 2}$ by using long division.

Solution

✎ Now You Try It!

Use the space provided to work out the solution to the next example.

Example A Using Long Division (Terms Missing)

Simplify $\dfrac{7x^3 + 4x - 9}{x - 2}$ by using long division.

Looking Ahead

Now that you have reviewed the division of polynomials, you will see how the same techniques can be used to tackle problems involving rational functions, as in the example below.

Example Preview

Find the equation(s) of the horizontal or oblique asymptotes of the following rational function.

$$h(x) = \frac{4x^2 + 6x - 6}{x + 5}$$

Solution

First, determine whether the rational function has horizontal or oblique asymptotes by comparing the degree of the polynomial in the numerator with the degree of the polynomial in the denominator. In this case, the degree of the numerator is 2 and the degree of the denominator is 1.

Since the degree of the numerator is equal to 1 plus the degree of the denominator, the line $y = g(x)$ is an oblique asymptote for h, where g is the quotient of the numerator and denominator.

$$\begin{array}{r}
4x - 14 \\
x+5 \overline{\smash{)}\, 4x^2 + 6x - 6} \\
\underline{-(4x^2 + 20x)} \\
-14x - 6 \\
\underline{-(-14x - 70)} \\
64
\end{array}$$

Thus, the oblique asymptote is $y = 4x - 14$.

5.R.4 Exercises

Concept Check

True/False. Determine whether each statement is true or false. If a statement is false, explain how it can be changed so the statement will be true. (**Note:** There may be more than one acceptable change.)

1. When dividing polynomials, any remainder must be of smaller degree than the divisor.

2. The first step in the division algorithm is to align the polynomials in ascending order.

3. To aid in organization and clarity when dividing polynomials, it is best to fill in any missing powers with ones.

4. The process followed when dividing two polynomials is called the division algorithm with polynomials.

Practice

Express each quotient as a sum (or difference) of fractions and simplify, if possible.

5. $\dfrac{8y^3 - 16y^2 + 24y}{8y}$

6. $\dfrac{20y^5 - 14y^4 + 21y^3 + 42y^2}{4y^2}$

Divide by using the division algorithm. Write the answers in the form $Q + \dfrac{R}{D}$, where the degree of R < the degree of D.

7. $\dfrac{x^2 - 2x - 20}{x + 4}$

8. $\dfrac{21x^3 + 41x^2 + 13x + 5}{3x + 5}$

9. $\dfrac{x^4 - 3x^3 + 2x^2 - x + 2}{x - 3}$

10. $\dfrac{x^3 - 27}{x - 3}$

Applications

Solve.

11. **Geometry:** A moving company uses a box that has a volume of $x^3 - 2x^2 - 13x - 10$ cubic inches.

    a. If the height of the box is $x + 2$, what is the area of the base of the box?

    b. If the height of the box is $x + 1$, what is the area of the base of the box?

Writing & Thinking

12. Suppose that a polynomial is divided by $(3x-2)$ and the answer is given as $x^2 + 2x + 4 + \dfrac{20}{3x-2}$. What was the original polynomial? Explain how you arrived at this conclusion.

5.R.5 Introduction to Rational Expressions

↻ Making Connections

A core concept of working with rational functions is determining their domain and range. You will find values that are excluded from the rational function's domain, in the form of holes and vertical asymptotes on the graph, by finding values that make the denominator equal to zero. Reviewing how to simplify rational expressions will assist you in finding the values to be excluded from the rational function's domain and identifying vertical asymptotes.

In this section, you will learn skills that you can apply when answering questions like these:

Find the equations of the vertical asymptotes, if any, for each of the following rational functions.

- $f(x) = \dfrac{5}{x-1}$
- $f(x) = \dfrac{x^2-4}{x+2}$

✖ Building Foundations

Rational Expressions

A rational expression is an algebraic expression that can be written in the form

_____

DEFINITION

Summary of Arithmetic Rules for Rational Numbers (or Fractions)

A **fraction** (or **rational number**) is a number that can be written in the form _____

_____

The Fundamental Principle: _____

The **reciprocal** of $\dfrac{a}{b}$ is _____

Multiplication: _____

Division: _____

Addition: _____

Subtraction: _____

PROPERTIES

The Fundamental Principle of Rational Expressions

If $\dfrac{P}{Q}$ is a rational expression and P, Q, and K are polynomials where

$$Q, K \neq 0, \text{ then}$$

_____

DEFINITION

▶ Watch and Work

Watch the video for Example 3 in the software and follow along in the space provided.

Example 3 Reducing Rational Expressions

Use the fundamental principle to reduce each expression to lowest terms. State any restrictions on the variable by using the fact that no denominator can be 0. This restriction applies to denominators **before and after** a rational expression is reduced.

a. $\dfrac{2x-10}{3x-15}$

b. $\dfrac{x^2-3x-4}{x^2-16}$

c. $\dfrac{y-10}{10-y}$

Solution

✏️ Now You Try It!

Use the space provided to work out the solution to the next example.

Example A Reducing Rational Expressions

Reduce each expression to lowest terms. State any restrictions on the variable.

a. $\dfrac{2x-6}{5x-15}$

b. $\dfrac{x^2 - x - 20}{x^2 - 25}$

c. $\dfrac{x-5}{5-x}$

Opposites in Rational Expressions

For a polynomial P, _____

In particular, _____

DEFINITION

5.R.5   Introduction to Rational Expressions

🔭 Looking Ahead

Your review of rational expressions and how to simplify them will help you with identifying the vertical asymptotes, if they exist, of a given rational function.

Example Preview

Find equations for the vertical asymptotes, if any, for the following rational function.

$$f(x) = \frac{-6x^2 + 4x + 2}{-2x + 2}$$

Solution

First, we need to reduce the function by removing any common factors in the numerator and denominator.

$$f(x) = \frac{-6x^2 + 4x + 2}{-2x + 2}$$
$$= \frac{\cancel{(-2x+2)}(3x+1)}{\cancel{(-2x+2)}}$$
$$= 3x + 1$$

Now, the vertical asymptotes of f are at the zeros of the denominator.

Since no denominator remains after canceling the common factors, there are no vertical asymptotes for f. (Note that the domain of f excludes $x = 1$. This means the graph has a "hole" at this x-value instead of a vertical asymptote.)

5.R.5 Exercises

Concept Check

True/False. Determine whether each statement is true or false. If a statement is false, explain how it can be changed so the statement will be true. (**Note:** There may be more than one acceptable change.)

1. A simplified rational expression cannot have any common factors other than 1 and −1 in both the numerator and denominator.

2. The difference between a rational number and a rational expression is that a rational expression generally has polynomials in the numerator and/or denominator.

3. While a rational number cannot have a zero denominator, a rational expression can have a zero denominator.

4. If a denominator is $x + 5$, it is defined for all values except 5.

Practice

Reduce each expression to lowest terms. State any restrictions on the variable(s).

5. $\dfrac{9x^2y^3}{12xy^4}$

6. $\dfrac{2x-8}{16-4x}$

7. $\dfrac{xy-3y+2x-6}{y^2-4}$

8. $\dfrac{x^2+10x+24}{2x^2+x-28}$

9. $\dfrac{x}{x^2-3x}$

10. Evaluate $\dfrac{3y-4}{y^2+25}$ for $y = 3$

Applications

Solve.

11. **Event Planning:** The cost of renting a party room with tables, chairs, and simple decorations is $200 plus $15 per person attending.

    a. Write a rational expression that represents the total price per person for renting the party room, where x is the number of people attending.

    b. What is the price per person to rent the party room if 10 people are attending?

    c. Determine which values of the variable will make the rational expression from Part **a.** undefined.

    d. Considering the context of the given problem, are there any additional restrictions on the variable? If so, explain why these restrictions are in place.

12. **Rectangles:** The area of a rectangle (in square feet) is represented by the polynomial function $A(x) = 4x^2 - 4x - 15$. If the length of the rectangle is $(2x+3)$ feet, find a representation for the width.

    $A(x) = 4x^2 - 4x - 15$

    $2x + 3$

Writing & Thinking

13. **a.** Define the term rational expression.

    b. Give an example of a rational expression that is undefined for $x = -2$ and $x = 3$ and has a value of 0 for $x = 1$. Explain how you determined this expression.

    c. Give an example of a rational expression that is undefined for $x = -5$ and never has a value of 0. Explain how you determined this expression.

5.R.6 Multiplication and Division with Rational Expressions

⟳ Making Connections

Understanding the division and multiplication of rational expressions is an important skill that we can apply to functions that are written as quotients of other functions.

In this section, you will learn skills that you can apply when answering questions like these:

- Find a formula for $\left(\dfrac{f}{g}\right)(x)$ for the following functions.

$$f(x) = \dfrac{x}{x-3} \quad \text{and} \quad g(x) = 5x$$

- Find a formula for $(fg)(x)$ for the following functions.

$$f(x) = \dfrac{x^2 - 1}{x} \quad \text{and} \quad g(x) = \dfrac{x}{x+1}$$

🛠 Building Foundations

> **To Multiply Rational Expressions**
>
> To multiply any two (or more) rational expressions,
>
> 1. completely factor each _____
>
> 2. multiply the numerators and _____
>   _____
>
> 3. "divide out" any common factors from the _____
>   _____
>
> PROCEDURE

> **Multiplying Rational Expressions**
>
> If $P, Q, R,$ and S are polynomials and $Q, S \neq 0$, then
>
> _____
>
> DEFINITION

Dividing Rational Expressions

If $P, Q, R,$ and S are polynomials with $Q, R, S \neq 0,$ then

$$\underline{}$$

Note that $\dfrac{S}{R}$ is the $\underline{}$

DEFINITION

▶ Watch and Work

Watch the video for Example 8 in the software and follow along in the space provided.

Example 8 Dividing with Rational Expressions

Divide and reduce, if possible. Assume that no denominator has a value of 0.

$$\frac{x^2 - 8x + 15}{2x^2 + 11x + 5} \div \frac{2x^2 - 5x - 3}{4x^2 - 1}$$

Solution

✏ Now You Try It!

Use the space provided to work out the solution to the next example.

Example A Dividing with Rational Expressions

Divide and reduce, if possible. Assume that no denominator has a value of 0.

$$\frac{x^2 - 9x + 18}{3x^2 + 19x + 6} \div \frac{3x^2 - 17x - 6}{x^2 + x - 30}$$

👁 Looking Ahead

Now that you have reviewed the main operations involving rational expressions, you will be able to apply these skills to combining functions. The following example illustrates combining functions using division.

Example Preview

Find a formula for $\left(\dfrac{f}{g}\right)(x)$ for the following functions.

$$f(x) = \frac{x^2 - 25}{x} \quad \text{and} \quad g(x) = \frac{x+5}{2x}$$

Solution

$$\left(\frac{f}{g}\right)(x) = \frac{f(x)}{g(x)}$$

$$= \frac{\dfrac{x^2-25}{x}}{\dfrac{x+5}{2x}}$$

$$= \frac{x^2-25}{x} \cdot \frac{2x}{x+5}$$

$$= \frac{\cancel{(x+5)}(x-5)\,2\cancel{x}}{\cancel{x}\cancel{(x+5)}}$$

$$= \frac{2(x-5)}{1}$$

$$= 2x-10$$

5.R.6 Exercises

Concept Check

True/False. Determine whether each statement is true or false. If a statement is false, explain how it can be changed so the statement will be true. (**Note:** There may be more than one acceptable change.)

1. The reciprocal of $\dfrac{x}{x+3}$ is $\dfrac{-x-3}{x}$.

2. Dividing rational expressions is similar to dividing fractions.

3. There are no restrictions on the denominator $12x^2$.

4. Because $\dfrac{4x^2}{16x}$ reduces to $\dfrac{x}{4}$, there are no restrictions on the denominator.

Practice

Perform the indicated operations and reduce to lowest terms. Assume that no denominator has a value of 0.

5. $\dfrac{x^2-9}{x^2+2x} \cdot \dfrac{x+2}{x-3}$

6. $\dfrac{2x^2+x-3}{x^2+4x} \cdot \dfrac{2x+8}{x-1}$

7. $\dfrac{x-1}{6x+6} \div \dfrac{2x-2}{x^2+x}$

8. $\dfrac{x+3}{x^2+3x-4} \div \dfrac{x+2}{x^2+x-2}$

Applications

Solve

9. **Carpentry:** Erik is building a cubby bookshelf, that is, a bookshelf divided into storage holes (cubbies) instead of shelves. He wants the height of the bookshelf to be $x^2 - 3x - 10$ and the width to be $x^2 + 5x + 6$. Each cubby hole in the bookshelf will have a height of $x + 3$ and a width of $x - 5$.

   a. Write a rational expression to determine how many cubbies high the bookshelf will be.

   b. Write a rational expression to determine how many cubbies wide the bookshelf will be.

   c. Multiply the rational expressions from Parts **a.** and **b.** (and reduce to lowest terms) to obtain a rational expression that gives the total number of cubbies in the entire bookshelf.

5.R.7 Simplifying Complex Fractions

♻ Making Connections

The ability to simplify complex rational expression allows us to reduce seemingly complicated expressions to more manageable ones.

In this section, you will learn skills that you can apply when answering questions like these:

- Find a formula for $\left(\dfrac{f}{g}\right)(x)$ for the following functions.

$$f(x) = \dfrac{x+3}{x} \quad \text{and} \quad g(x) = \dfrac{1}{x}$$

- Find a formula for both $(f \circ g)(x)$ and $(g \circ f)(x)$ for the following functions.

$$f(x) = \dfrac{x-4}{x} \quad \text{and} \quad g(x) = \dfrac{1}{x+1}$$

🛠 Building Foundations

> **To Simplify Complex Fractions (First Method)**
> 1. Simplify the numerator so that _____
> 2. Simplify the denominator so that _____
> 3. Divide the _____
>
> PROCEDURE

> **To Simplify Complex Fractions (Second Method)**
> 1. Find the LCM of all the denominators in _____
> 2. Multiply both the numerator and denominator of _____
> 3. Simplify both the numerator and _____
>
> PROCEDURE

▶ Watch and Work

Watch the video for Example 4 in the software and follow along in the space provided.

Example 4 Second Method for Simplifying Complex Fractions

Simplify the complex fraction. $\dfrac{\dfrac{1}{x+3} - \dfrac{1}{x}}{1 + \dfrac{3}{x}}$

Solution

✏ Now You Try It!

Use the space provided to work out the solution to the next example.

Example A Second Method for Simplifying Complex Fractions

Simplify the complex fraction. $\dfrac{\dfrac{1}{x+6} - \dfrac{1}{x}}{1 + \dfrac{6}{x}}$

Looking Ahead

Now that you have reviewed how to simplify complex rational expressions, you will be able to apply these skills to more advanced topics like the composition of rational functions. The following example shows you how to apply the skills of this section to such a problem.

Example Preview

Find the formula for $(g \circ f)(x)$ for the following functions.

$$f(x) = \frac{1}{x} \quad \text{and} \quad g(x) = \frac{x-2}{5}$$

Solution

$$(g \circ f)(x) = g(f(x))$$
$$= g\left(\frac{1}{x}\right)$$
$$= \frac{\left(\frac{1}{x}\right) - 2}{5}$$
$$= \frac{\frac{1}{x} - 2}{5} \cdot \frac{x}{x}$$
$$= \frac{1 - 2x}{5x}$$

5.R.7 Exercises

Concept Check

True/False. Determine whether each statement is true or false. If a statement is false, explain how it can be changed so the statement will be true. (**Note:** There may be more than one acceptable change.)

1. When simplifying complex fractions, the answer should always be reduced to lowest terms.

2. Complex fractions are those fractions in which only the denominator consists of one or more fractions itself.

5.R.7 Simplifying Complex Fractions

3. Sometimes finding the LCM of all denominators is an important first step for simplifying complex fractions.

4. The LCM of the denominators of $\dfrac{2}{x-6}$ and $\dfrac{x}{6}$ is 6.

Practice

Simplify the following complex fractions.

5. $\dfrac{\dfrac{2x}{3y^2}}{\dfrac{5x^2}{6y}}$

6. $\dfrac{\dfrac{x+3}{2x}}{\dfrac{2x-1}{4x^2}}$

7. $\dfrac{\dfrac{3}{x}+\dfrac{5}{2x}}{\dfrac{1}{x}+4}$

8. $\dfrac{\dfrac{7}{x}-\dfrac{14}{x^2}}{\dfrac{1}{x}-\dfrac{4}{x^3}}$

Simplify the following complex algebraic expressions.

9. $\dfrac{1}{x+1} - \dfrac{3}{2x} \cdot \dfrac{4x}{x+1}$

10. $\dfrac{x}{x-1} - \dfrac{3}{x-1} \cdot \dfrac{x+2}{x}$

Applications
Solve.

11. *Investing:* The average percent yield (APY) of an annuity is the annual interest rate earned in a given year that accounts for the effects of compounding. The APY acts as the interest rate for a simple interest account and is larger than the stated interest rate on the compound interest account. The formula to calculate the APY on an annuity after 2 years is

$$\text{APY} = \left(1 + \dfrac{r}{2}\right)^2 - 1,$$

where r is the stated interest rate.

  a. Simplify the expression for APY and write as a single rational expression.

  b. Using the original formula, calculate the APY for an annuity whose interest rate is 6%. Do not round.

  c. Using the expression in Part a., calculate the APY for an annuity whose interest rate is 6%. Do not round.

  d. Does the result from Part c. match the result from Part b.? Explain why or why not.

  e. How much larger is the APY than the interest rate?

  f. Why do you think the APY is larger than the interest rate? Write a complete sentence.

Writing & Thinking

12. Some complex fractions involve the sum (or difference) of complex fractions. Beginning with the outermost denominator, simplify each of the following expressions.

    a. $1 + \dfrac{1}{1 + \dfrac{1}{1 + \dfrac{1}{1+1}}}$

    b. $2 - \dfrac{1}{2 - \dfrac{1}{2 - \dfrac{1}{2-1}}}$

    c. $x + \dfrac{1}{x + \dfrac{1}{x + \dfrac{1}{x+1}}}$

CHAPTER 7.R

Review Concepts

for *Exponential and Logarithmic Functions*

7.R.1 Rules for Exponents

7.R.2 Power Rules for Exponents

7.R.3 Rational Exponents

7.R.4 Introduction to Logarithmic Functions

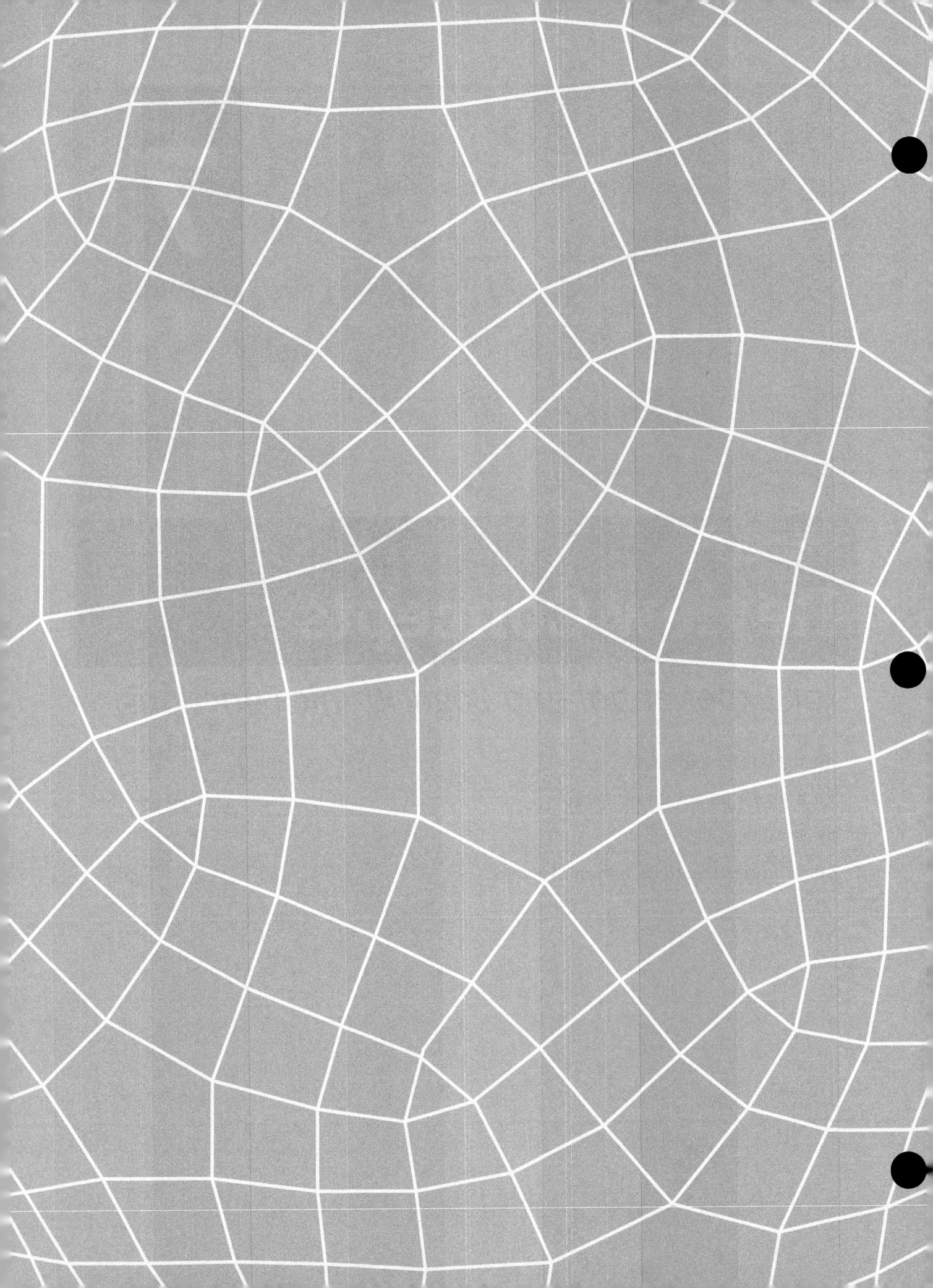

7.R.1 Rules for Exponents

⟳ Making Connections

When graphing exponential functions and solving exponential equations, you will often have to simplify expressions with integer exponents. Therefore, it is important to review how to use the product and quotient rules for exponents, as well as how to handle negative exponents and an exponent of zero.

In this section, you will learn skills that you can apply when answering questions like these:

- Sketch the graph of the function $p(x) = 3^{x-2}$.

- Solve the exponential equation $2^x = \left(\dfrac{1}{2}\right)^{13}$.

🛠 Building Foundations

The Product Rule for Exponents

If a is a nonzero real number and m and n are integers, then

_____

In words, to multiply powers with the same base, _____

PROPERTIES

The Exponent 0

If a is a nonzero real number, then _____

The expression _____ is _____

DEFINITION

Quotient Rule for Exponents

If a is a nonzero real number and m and n are integers, then

_____

In words, to divide two powers with the same base, _____

PROPERTIES

7.R.1 Rules for Exponents

Rule for Negative Exponents

If a is a nonzero real number and n is an integer, then

_____

PROPERTIES

▶ Watch and Work

Watch the video for Example 6 in the software and follow along in the space provided.

Example 6 Negative Exponents

Use the rule for negative exponents to simplify each expression so that it contains only positive exponents.

a. 5^{-1}

b. x^{-3}

c. $x^{-9} \cdot x^{7}$

Solution

✏️ Now You Try It!

Use the space provided to work out the solution to the next example.

Example A Negative Exponents

Use the rule for negative exponents to simplify each expression so that it contains only positive exponents.

a. 7^{-1}

b. x^{-7}

c. $x^{-11} \cdot x^6$

Summary of the Rules for Exponents

For any nonzero real number a and integers m and n:

1. The exponent 1: _____

2. The exponent 0: _____

3. The product rule: _____

4. The quotient rule: _____

5. Negative exponents: _____

PROPERTIES

Looking Ahead

The following example requires you to use the skills you learned to rewrite a number with a different base using negative exponents. Once the expressions on opposite sides of the equation have the same base, they can only be equal if their exponents are equal. This leads to a simple linear equation in one variable that can easily be solved.

Example Preview

Solve the following exponential equation.

$$3^{3x-5} = \frac{1}{9}$$

Solution

Solving this exponential equation involves rewriting the terms on both sides of the equation with the same base. Once this is done, the exponents can be equated, and the subsequent equation can be solved for x.

This exponential equation can be solved in the following manner.

$$3^{3x-5} = \frac{1}{9}$$

$$3^{3x-5} = \left(\frac{1}{3}\right)^2$$

$$3^{3x-5} = 3^{-2}$$

$$3x - 5 = -2 \quad \text{Exponents must be equal with the same base.}$$

$$3x = 3$$

$$x = 1$$

7.R.1 Exercises

Concept Check

True/False. Determine whether each statement is true or false. If a statement is false, explain how it can be changed so the statement will be true. (**Note:** There may be more than one acceptable change.)

1. If a constant does not have an exponent written, it is assumed that the exponent is 0.

2. If a is a nonzero real number and n is an integer, then $a^{-n} = -a^n$.

3. Since the product rule is stated for integer exponents, the rule is also valid for 0 and negative exponents.

4. When using the quotient rule, you should subtract the smaller exponent from the larger exponent.

Practice

Simplify each expression. The final form of the expressions with variables should contain only positive exponents. Assume that all variables represent nonzero numbers.

5. $y^3 \cdot y^8$

6. $\dfrac{y^7}{y^2}$

7. $x^{-3} \cdot x^0 \cdot x^2$

8. $\dfrac{10^4 \cdot 10^{-3}}{10^{-2}}$

9. $(9x^2 y^3)(-2x^3 y^4)$

10. $\dfrac{-8x^{-2} y^4}{4x^2 y^{-2}}$

Applications

Solve.

11. *Computers:* Rylee wants to move all her files to a new hard drive that has 2^{12} GB of storage on it. She wants to designate the same amount of storage for each of 2^4 projects. How much storage should be assigned to each project? Write your answer as a power of two.

12. **Bacteria:** Trey is studying patterns in bacteria. For a positive test result in his experiment, bacteria must grow in population at a minimum rate of 3^2 in 24 hours. If the initial population of the bacteria is 3^5 and his final measurement after 24 hours is 3^8, should he mark the test as positive or negative?

7.R.2 Power Rules for Exponents

↻ Making Connections

When graphing exponential functions and solving exponential equations, you will often have to simplify expressions with integer exponents. Therefore, it is important to review the power rule for exponents, the rule for power of a product, and the rule for power of a quotient.

In this section, you will learn skills that you can apply when answering questions like these:

- Sketch the graph of the function $p(x) = \left(\dfrac{9}{2}\right)^{-x}$.
- Solve the following exponential equation $9^{2x-5} = 27^{x-2}$.

🛠 Building Foundations

Power Rule for Exponents

If a is a nonzero real number and m and n are integers, then

_____

In other words, the value of a power raised to a power can be found by _____

PROPERTIES

Rule for Power of a Product

If a and b are nonzero real numbers and n is an integer then

_____

In words, a power of a product is found by _____

PROPERTIES

Rule for Power of a Quotient

If a and b are nonzero real numbers and n is an integer, then

_____

In words, a power of a quotient (in fraction form) is found by _____

PROPERTIES

▶ Watch and Work

Watch the video for Example 5 in the software and follow along in the space provided.

Example 5 Using Two Approaches with Fractional Expressions and Negative Exponents

Simplify: $\left(\dfrac{x^3}{y^5}\right)^{-4}$

Solution

✏ Now You Try It!

Example A Using Two Approaches with Fractional Expressions and Negative Exponents

Simplify:

$\left(\dfrac{x^6}{y^3}\right)^{-5}$

Summary of the Rules for Exponents

For any nonzero real numbers a and b and integers m and n:

1. The exponent 1: _____

2. The exponent 0: _____

3. The product rule: _____

4. The quotient rule: _____

5. Negative exponents: _____

6. Power rule: _____

7. Power of a product: _____

8. Power of a quotient: _____

PROPERTIES

👁 Looking Ahead

The following example incorporates many of the skills you have reviewed including the power rule for exponents. Once the expressions on opposite sides of the equation have the same base, they can only be equal if their exponents are equal. This leads to a simple linear equation in one variable that can easily be solved.

Example Preview

Solve the following exponential equation.

$$1000^{-x} = 10^{4x-5}$$

Solution

Solving this exponential equation involves rewriting the terms on both sides of the equation with the same base. Once this is done, the exponents can be equated, and the subsequent equation can be solved for x.

This exponential equation can be solved in the following manner.

$$\begin{aligned}
1000^{-x} &= 10^{4x-5} \\
\left(10^3\right)^{-x} &= 10^{4x-5} \\
10^{-3x} &= 10^{4x-5} \quad &&\text{Power Rule } (a^m)^n = a^{mn} \\
-3x &= 4x - 5 \quad &&\text{Exponents must be equal with the same base.} \\
-7x &= -5 \\
x &= \frac{5}{7}
\end{aligned}$$

7.R.2 Exercises

Concept Check

True/False. Determine whether each statement is true or false. If a statement is false, explain how it can be changed so the statement will be true. (**Note:** There may be more than one acceptable change.)

1. Taking the reciprocal of a fraction changes the sign of any exponent in the fraction.

2. For an exponent to refer to −7 as the base, −7 must be in parentheses.

3. When simplifying an expression with exponents, the rules for exponents must be used in a specific order or the answer will vary.

4. The expression -8^2 simplifies to −64.

Practice

Use the rules for exponents to simplify each of the expressions. Assume that all variables represent nonzero real numbers.

5. $\left(2^{-3}\right)^{-2}$

6. $-3\left(7xy^2\right)^0$

7. $-2\left(3x^5 y^{-2}\right)^{-3}$

8. $\left(\dfrac{x}{2}\right)^3$

9. $\left(\dfrac{2x^2 y}{y^3}\right)^{-4}$

10. $\left(\dfrac{5a^4 b^{-2}}{6a^{-4} b^3}\right)^{-2} \left(\dfrac{5a^3 b^4}{2^{-2} a^{-2} b^{-2}}\right)^3$

7.R.3 Rational Exponents

↻ Making Connections

When graphing exponential functions and logarithmic functions, as well as evaluating exponential and logarithmic expressions, you will often have to simplify expressions with rational exponents or radical expressions. Therefore, it is important to review how to use the product rule, quotient rule, power rule, rule for power of a product, and rule for power of a quotient with rational exponents. It is also essential to review how to translate radical expressions into expressions with rational exponents and vice versa.

In this section, you will learn skills that you can apply when answering questions like these:

- Sketch the graph of the function $h(x) = 3^{\left(\frac{1}{2}\right)^x}$.

- Solve the exponential equation $\left(\frac{1}{5}\right)^{x-4} = 625^{\frac{1}{2}}$.

- Solve the logarithmic equation $\log_{16} x^{\frac{1}{2}} = \left(\frac{3}{4}\right)$.

🛠 Building Foundations

Radical Notation

If n is an integer greater than 1, then _____

The expression $\sqrt[n]{a}$ is called _____

The symbol $\sqrt[n]{}$ is called _____

n is called _____

a is called _____

Note: If no index is given, it is _____

For example, _____

DEFINITION

The General Form $a^{\frac{m}{n}}$

If n is an integer greater than 1, m is any integer, and $a^{\frac{1}{n}}$ is a real number, then

_____

In radical notation:

_____

DEFINITION

▶ Watch and Work

Watch the video for Example 4 in the software and follow along in the space provided.

Example 4 Simplifying Radical Notation by Changing to Exponential Notation

Simplify each expression by first changing it into an equivalent expression with rational exponents. Then rewrite the answer in simplified radical form. Assume that all variables represent positive real numbers.

a. $\sqrt[4]{\sqrt[3]{x}} =$

b. $\sqrt[3]{a}\sqrt{a} =$

c. $\dfrac{\sqrt{x^3}\,\sqrt[3]{x^2}}{\sqrt[5]{x^2}} =$

✏ Now You Try It!

Use the space provided to work out the solution to the next example.

Example A Simplifying Radical Notation by Changing to Exponential Notation

Rewrite the expression in simplified radical form. Assume that all variables represent positive real numbers.

a. $\sqrt[5]{\sqrt{x}}$

b. $\sqrt[4]{x} \cdot \sqrt{x}$

c. $\dfrac{\sqrt[3]{x^2} \cdot \sqrt{x^3}}{\sqrt[6]{x^5}}$

🔭 Looking Ahead

The following example incorporates many of the skills you have reviewed including the power rule with rational exponents, translating expressions with rational exponents to radicals, and evaluating radicals.

Example Preview

Use the elementary properties of logarithms to solve the following equation.

$$\log_{27}(x) = \frac{2}{3}$$

Solution

If we consider the definition of logarithmic functions, we know that the following two equations are equivalent.

$$x = a^y \text{ and } y = \log_a(x)$$

Therefore, this equation can be equivalently rewritten as $x = 27^{\frac{2}{3}}$.

$$x = (27)^{\frac{2}{3}}$$
$$x = \left(27^{\frac{1}{3}}\right)^2 \quad \text{Power Rule } (a^m)^n = a^{mn}$$
$$x = \left(\sqrt[3]{27}\right)^2 \quad a^{\frac{1}{3}} = \sqrt[3]{a}$$
$$x = 3^2$$
$$x = 9$$

7.R.3 Exercises

Concept Check

True/False. Determine whether each statement is true or false. If a statement is false, explain how it can be changed so the statement will be true. (**Note:** There may be more than one acceptable change.)

1. The same rules for exponents apply to both integer exponents and rational exponents.

2. If the cube root of 7 were to be converted into exponential notation it would be $\sqrt[3]{7}$.

3. Any expression to the power 0, such as $\left(\sqrt[4]{x}\right)^0$, is equal to 1.

4. The expression $y^{\frac{1}{2}}$ can be rewritten in radical notation as $\sqrt{y^2}$.

Practice

5. Use radical notation to write an expression that is equivalent to $8^{\frac{1}{3}}$.

6. Use exponential notation to write an expression that is equivalent to $\sqrt{3}$.

Simplify each numerical expression.

7. $100^{-\frac{1}{2}}$

8. $64^{\frac{2}{3}}$

9. Simplify $\dfrac{a^{\frac{1}{2}} \cdot a^{-\frac{3}{4}}}{a^{-\frac{1}{2}}}$. Assume that all variables represent positive real numbers.

10. Simplify $\dfrac{\sqrt[4]{y^3}}{\sqrt[6]{y}}$ by first changing it into an equivalent expression with rational exponents. Rewrite the answer in simplified radical form. Assume that all variables represent positive real numbers.

Applications
Solve.

11. *Area:* The width of a rectangle is $\sqrt[3]{64^2}$ ft and the length is $216^{\frac{2}{3}}$ ft. What is the area of the rectangle?

12. *Amusement Parks:* An amusement park is creating signs to indicate the velocity of the roller coaster car on certain hills of the most popular rides. A roller coaster car gains kinetic energy as it goes down a hill. The velocity, or speed, of an object in kilometers per hour (km/h) can be determined by $V = \left(\dfrac{2k}{m}\right)^{\frac{1}{2}}$, where k is the kinetic energy of the object in joules (J) and m is the mass of the object in kilograms (kg).

   a. For the most popular roller coaster, the car has a mass of 300 kg and the car has a kinetic energy of 375,000 J on the first hill. What velocity does the car obtain on the first hill?

   b. For the second most popular roller coaster, the car has a mass of 350 kg and the car has a kinetic energy of 70,000 on the first hill. What velocity does the car obtain on the first hill?

Writing & Thinking

13. Is $\sqrt[5]{a} \cdot \sqrt{a}$ the same as $\sqrt[5]{a^2}$? Explain why or why not.

14. Assume that x represents a positive real number. Describe what kind of number the exponent n must be for x^n to mean

    a. a product.

    b. a quotient.

    c. 1.

    d. a radical expression.

7.R.4 Introduction to Logarithmic Functions

↻ Making Connections

Many exponential equations like $2^{x-1} = 8$ or $\left(\dfrac{1}{3}\right)^x = 9$ can be solved using only the properties of exponents. As you encounter more elaborate problems involving exponents, it will be helpful for you to know about logarithmic functions and their properties. This knowledge will also help you to solve equations that contain logarithms.

In this section, you will learn skills that you can apply when answering questions like these:

- Solve the following exponential equation.

$$3e^{3x+4} = 65$$

- Solve the following logarithmic equation.

$$\log(x-1) + \log(x+2) = 1$$

🛠 Building Foundations

The inverse of an exponential function is a _____.

Definition of Logarithm (base b)

For $b > 0$ and $b \neq 1$,

_____ is read _____

DEFINITION

A logarithm is _____.

Basic Properties of Logarithms

For $b > 0$ and $b \neq 1$,

1. $\log_b 1 = 0$ Regardless of the base, the _____
2. $\log_b b = 1$ The logarithm of the base _____
3. $x = b^{\log_b x}$ For _____
4. $\log_b b^x = x$

PROPERTIES

▶ Watch and Work

Watch the video for Example 3 in the software and follow along in the space provided.

Example 3 Solving Logarithmic Equations

Solve by first changing the equation to exponential form: $\log_{16} x = \dfrac{3}{4}$

Solution

✏ Now You Try It!

Use the space provided to work out the solution to the next example.

Example A Solving Logarithmic Equations

Solve by first changing the equation to exponential form:

$$\log_8 x = \dfrac{2}{3}$$

🔭 Looking Ahead

Being familiar with the main properties of logarithms and logarithmic functions allows you to tackle logarithmic and exponential equations without the need of a calculator. In the example below, you will use many of those skills to determine the value of x in an exponential equation.

Example Preview

Solve the following exponential equation.

$$e^{2x} + 4e^x - 45 = 0$$

Solution

This equation is in quadratic form. First, solve for e^x by factoring. Then, solve for x by converting from exponential to logarithmic form.

$$(e^x - 5)(e^x + 9) = 0$$
$$e^x = 5, -9$$
$$x = \ln(5), \ln(-9)$$

Since $\ln(-9)$ is undefined, the only solution is $\ln(5)$.

7.R.4 Exercises

Concept Check

True/False. Determine whether each statement is true or false. If a statement is false, explain how it can be changed so that the statement will be true. (**Note:** There may be more than one acceptable change.)

1. Exponential functions of the form $y = b^x$ are one-to-one functions and have inverses.

2. The exponent of an exponential function is the base of its inverse logarithmic function.

3. Exponents are logarithms.

4. The logarithm of the base is always 1.

Practice

5. Express $7^2 = 49$ in logarithmic form.

6. Express $\log_5 125 = 3$ in exponential form.

Solve by first changing each equation to exponential form.

7. $\log_5 \dfrac{1}{125} = x$

8. $\log_x 121 = 2$

9. $\log_8 8^{3.7} = x$

Graph the function and its inverse on the same set of axes.

10. $f(x) = 2^x$

Writing & Thinking

11. Discuss, in your own words, the symmetrical relationship of the graphs of the two logarithmic functions $y = \log_{10} x$ and $y = -\log_{10} x$.

CHAPTER 8.R

Review Concepts

for Conic Sections

8.R.1 Special Products of Binomials

8.R.2 Special Factoring Techniques

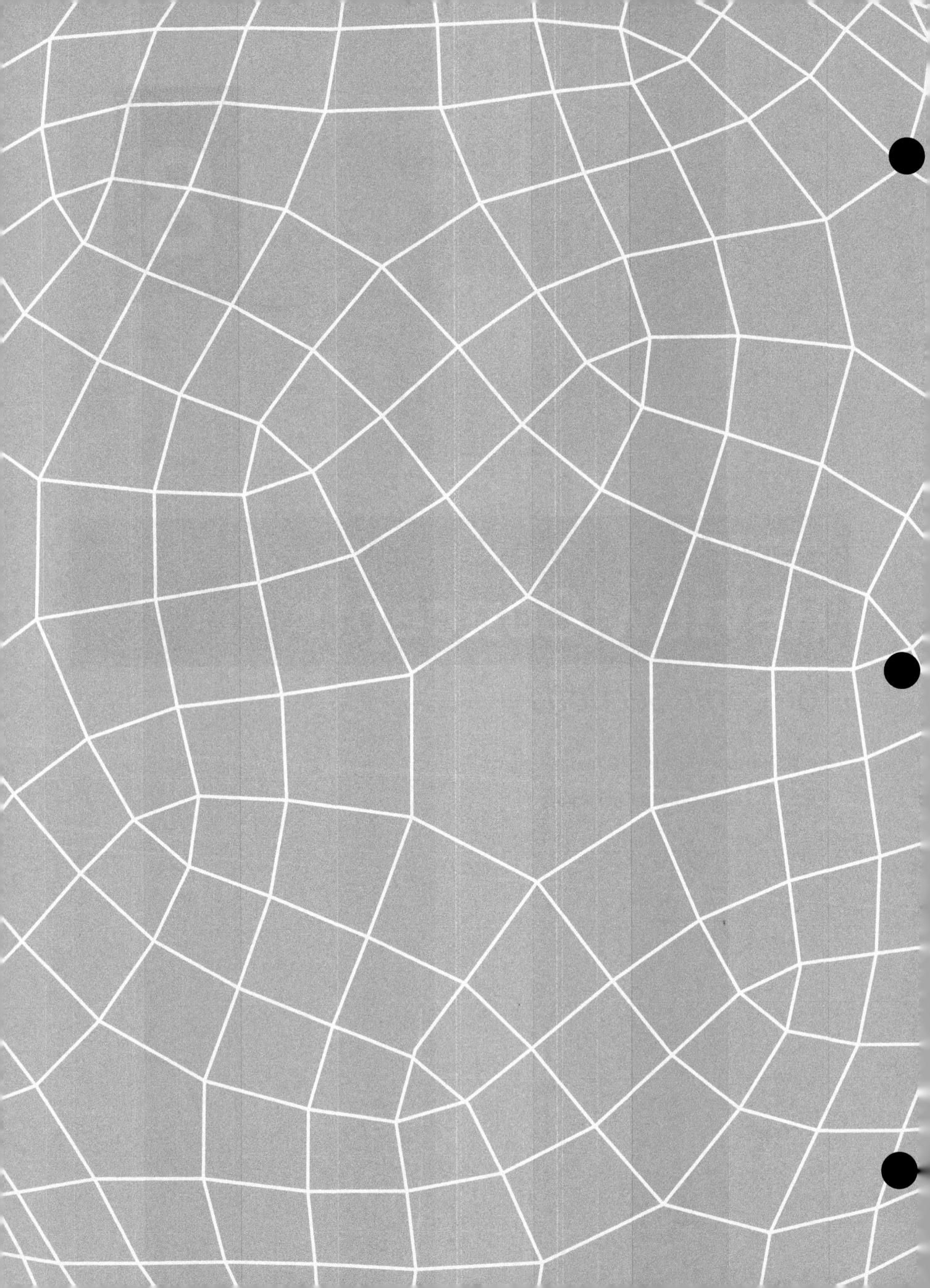

8.R.1 Special Products of Binomials

↻ Making Connections

The standard forms of each type of conic section contain special products of binomials. It is important to be able to recognize these special products if asked to write a conic section in standard form or if asked to find the equation for a conic section with known properties. Quickly calculating these products utilizing the rules associated with special products will be especially useful if you are asked to represent the conic section in the form $Ax^2 + Cy^2 + Dx + Ey + F = 0$ after being given its standard form.

In this section, you will learn skills that you can apply when answering questions like these:

- Find the center, foci, and vertices of the ellipse that the equation describes.

$$\frac{(x-5)^2}{4} + \frac{(y-2)^2}{25} = 1$$

- Find the equation of the following ellipse. Express your answer in standard form.

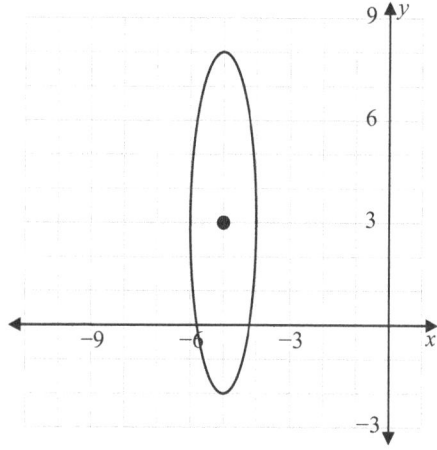

🛠 Building Foundations

Difference of Two Squares

_____

DEFINITION

Squares of Binomials (Perfect Square Trinomials)

$(x+a)^2 = x^2 + 2ax + a^2$ Square of a _____

$(x-a)^2 = x^2 - 2ax + a^2$ Square of a _____

DEFINITION

8.R.1 Special Products of Binomials

▶ Watch and Work

Watch the video for Example 3 in the software and follow along in the space provided.

Example 3 Squares of Binomials

Find the following products.

a. $(2x+3)^2$

b. $(5x-1)^2$

c. $(9-x)^2$

d. $(y^3+1)^2$

Solution

✏️ Now You Try It!

Use the space provided to work out the solution to the next example.

Example A Squares of Binomials

Find the following products.

a. $(3x+5)^2$

b. $(4x-2)^2$

c. $(8-2x)^2$

d. $(2y^3-2)^2$

👁 Looking Ahead

The following example illustrates the standard form of the equation for an ellipse, which contains the squares of two binomials.

Example Preview

Find the equation of the following ellipse.

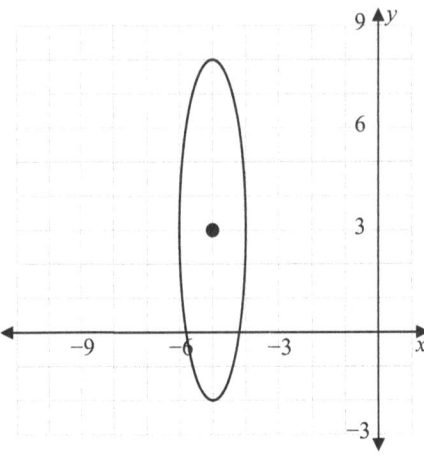

Solution

Since this ellipse is elongated vertically, the major axis is vertical and the standard form for the equation of this ellipse must be

$$\frac{(x-h)^2}{b^2} + \frac{(y-k)^2}{a^2} = 1$$

where the ellipse is centered at (h,k), the major axis has length $2a$, and the minor axis has length $2b$.

Careful examination of this graph shows that the center of this ellipse is located at $(-5,3)$. In addition, it can also be seen that the length of the major axis is 10 and the length of the minor axis is 2, which means that $a = 5$ and $b = 1$.

Substituting each of these values into the standard form of the equation results in the following equation of an ellipse.

$$(x+5)^2 + \frac{(y-3)^2}{25} = 1$$

8.R.1 Exercises

Concept Check

True/False. Determine whether each statement is true or false. If a statement is false, explain how it can be changed so the statement will be true. (**Note:** There may be more than one acceptable change.)

1. When two binomials are in the form of the sum and difference of the same term, the product will be a trinomial.

2. When the two binomials being multiplied together are the same, the product will be a trinomial.

3. Perfect square trinomials result from squaring a binomial sum or a binomial difference.

4. When finding the product of two binomials that are in the form of the sum and difference of the same two terms, the FOIL method and the difference of two squares formula will produce different results.

Practice

Find each product and identify any that are either the difference of two squares or a perfect square trinomial.

5. $(x-7)^2$

6. $(x+12)(x-12)$

7. $(x+4)(x+4)$

8. $(3x+7)^2$

9. $(3x-2)(3x-2)$

10. $(5x-9)(5x+9)$

Applications

Solve.

11. **Geometry:** A square is 20 inches on each side. A square x inches on each side is cut from each corner of the square.

    a. Represent the area of the remaining portion of the square in the form of a polynomial function $A(x)$.

    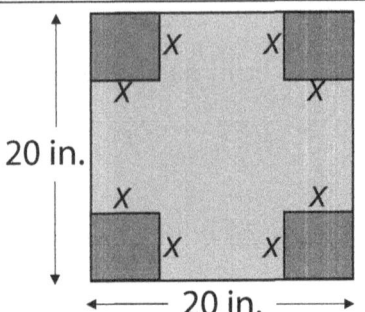

    b. Represent the perimeter of the remaining portion of the square in the form of a polynomial function $P(x)$.

12. **Probability:** In the case of binomial probabilities, if x is the probability of success in one trial of an event, then the expression $f(x) = 15x^4(1-x)^2$ is the probability of 4 successes in 6 trials where $0 \leq x \leq 1$.

    a. Represent the expression $f(x)$ as a single polynomial by multiplying the polynomials.

    b. If a fair coin is tossed, the probability of heads occurring is $\frac{1}{2}$. That is, $x = \frac{1}{2}$. Find the probability of 4 heads occurring in 6 tosses.

Writing & Thinking

13. A square with sides of length $(x+5)$ can be broken up as shown in the diagram. The sums of the areas of the interior rectangles and squares is equal to the total area of the square: $(x+5)^2$. Show how this fits with the formula for the square of a sum.

8.R.2 Special Factoring Techniques

↻ Making Connections

While the most useful form of a conic section is its standard form, you're often presented the conic section in the form $Ax^2 + Cy^2 + Dx + Ey + F = 0$. You must then manipulate the equation into one of the standard forms utilizing special factorizations. Reviewing how to recognize the special factorizations of squares and how to factor these special factorizations will make your work with conic sections easier to accomplish.

In this section, you will learn skills that you can apply when answering questions like these:

- Sketch the graph of the following ellipse and determine the coordinates of the foci.

$$4x^2 + 9y^2 + 40x + 90y + 289 = 0$$

- Sketch the graph of the following hyperbola, using asymptotes as guides.

$$9x^2 - 16y^2 + 116 = 36x + 64y$$

🛠 Building Foundations

Difference of Two Squares

Consider the polynomial $x^2 - 25$. By recognizing this expression as the **difference of two squares**, we can go directly to the factors:

$$x^2 - 25 = \underline{\hspace{4cm}}$$

DEFINITION

Sum of Two Squares

The sum of two squares is an expression of the form _____

DEFINITION

In a perfect square trinomial, both the _____ and _____ terms of the trinomial must be perfect squares. If the first term is of the form x^2 and the last term is of the form a^2, then the middle term must be of the form _____ or _____.

Watch and Work

Watch the video for Example 3 in the software and follow along in the space provided.

Example 3 Factoring Perfect Square Trinomials

Factor completely.

a. $z^2 - 12z + 36$

b. $4y^2 + 12y + 9$

c. $2x^3 - 8x^2y + 8xy^2$

d. $(x^2 + 6x + 9) - y^2$

Solution

✏ Now You Try It!

Use the space provided to work out the solution to the next example.

Example A Factoring Perfect Square Trinomials

Factor completely.

a. $z^2 + 40z + 400$

b. $y^2 - 14y + 49$

c. $3x^2z - 18xyz + 27y^2z$

d. $(y^2 + 8y + 16) - z^2$

8.R.2 Special Factoring Techniques

👀 Looking Ahead

The following example uses what you reviewed on the special factorizations of squares and the method of completing the square for a variable to write the equation for an ellipse in standard form and identify its center.

Example Preview

Consider the following equation of an ellipse.

$$49x^2 + 9y^2 - 72y - 297 = 0$$

Step 1: Rewrite this equation in the standard form of an ellipse.

Step 2: Find the center of the ellipse.

Solution

Step 1:

$$49x^2 + 9y^2 - 72y - 297 = 0$$
$$49x^2 + 9(y^2 - 8y) = 297$$
$$49x^2 + 9(y^2 - 8y + 16) = 297 + 144 \quad \text{Using the method of completing the square for } y.$$
$$49x^2 + 9(y-4)^2 = 441$$
$$\frac{x^2}{9} + \frac{(y-4)^2}{49} = 1$$

Step 2:

If we consider the standard form of an ellipse, it should be clear that, assuming a and b are fixed positive numbers with $a < b$, the standard form for the equation of this ellipse must be:

$$\frac{(x-h)^2}{b^2} + \frac{(y-k)^2}{a^2} = 1$$

Further, since the center of the ellipse is located at (h,k) the center of this ellipse is located at $(0,4)$.

8.R.2 Exercises

Concept Check

True/False. Determine whether each statement is true or false. If a statement is false, explain how it can be changed so the statement will be true. (**Note:** There may be more than one acceptable change.)

1. The expression $x^2 + 20x + 100$ is a perfect square trinomial.

2. When factoring polynomials, always look for a common monomial factor first.

3. The sum of two squares, $(x^2 + a^2)$, is factorable.

Practice

Completely factor each of the given polynomials. If a polynomial cannot be factored, write "not factorable."

4. $25 - z^2$

5. $y^2 - 16y + 64$

6. $x^2 + 64y^2$

7. $2x^2 - 128$

8. $25x^2 + 30x + 9$

9. $9x^2 - y^2$

Solve.

10. **a.** Represent the area of the shaded region of the square shown as the difference of two squares.

    b. Use the factors of the expression in Part **a.** to draw (and label the sides of) a rectangle that has the same area as the shaded region.

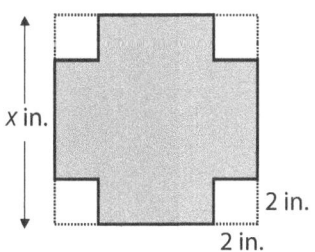

x in.

2 in.

2 in.

288   8.R.2   Special Factoring Techniques

Writing & Thinking

11. **a.** Show that the sum of the areas of the rectangles and squares in the figure is a perfect square trinomial.

   b. Rearrange the rectangles and squares in the form of a square and represent its area as the square of a binomial.

   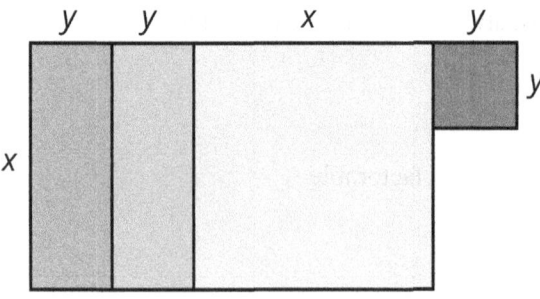

CHAPTER 9.R

Review Concepts

for *Systems of Equations and Inequalities*

9.R.1 Systems of Linear Equations: Solutions by Graphing

9.R.2 Systems of Linear Equations: Solutions by Substitution

9.R.3 Systems of Linear Equations: Solutions by Addition

9.R.4 Systems of Linear Inequalities

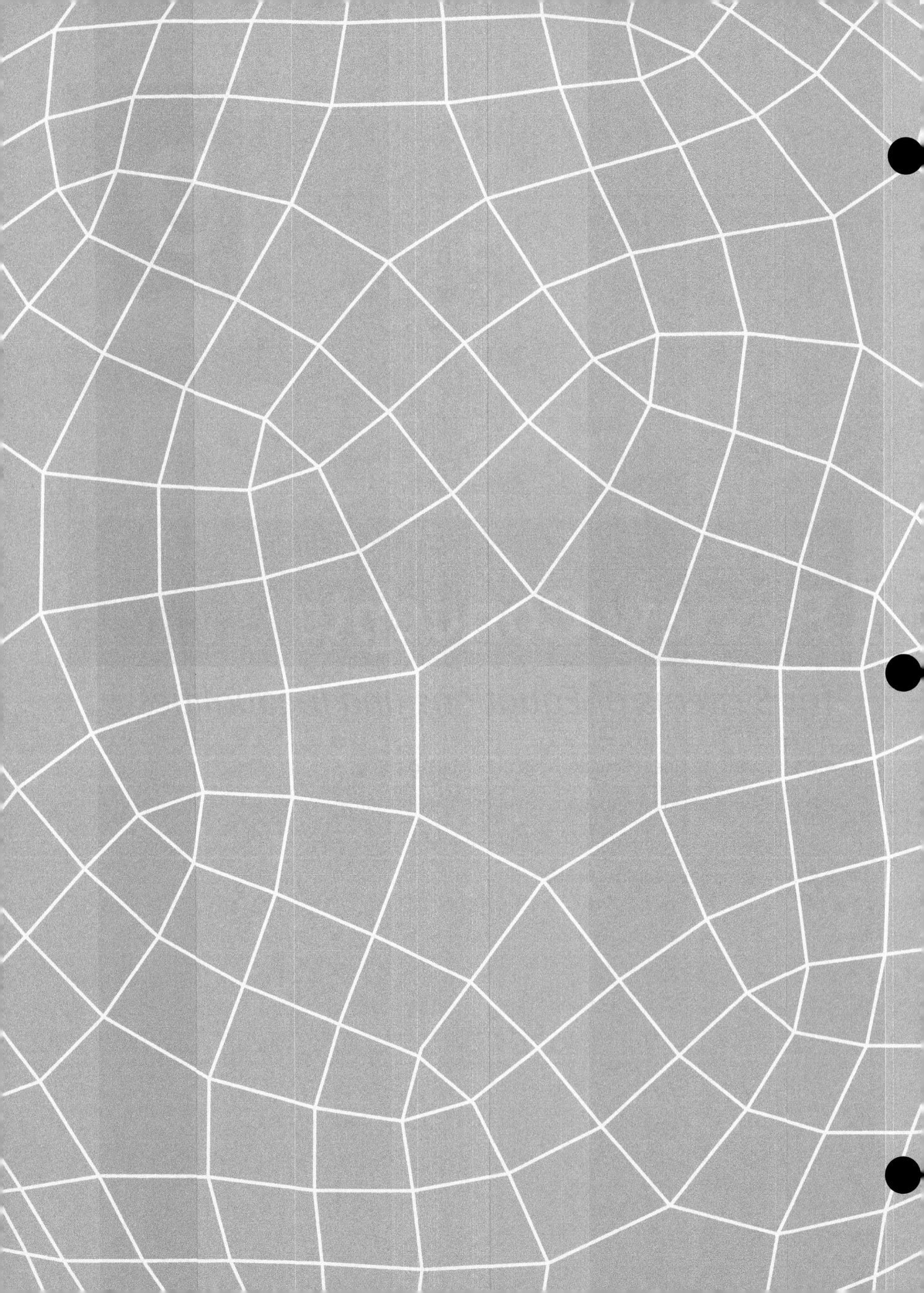

9.R.1 Systems of Linear Equations: Solutions by Graphing

⟲ Making Connections

Many real world applications involve using two variables that are associated through two linear equations. Graphing the corresponding system of two equations is helpful in visualizing the relationships between the two variables. Therefore, it is important to review how to solve a system of linear equations by graphing both equations and determining if there is a point of intersection that satisfies both equations.

In this section, you will learn skills that you can apply when answering questions like these:

- Solve the following system of equations.

$$\begin{cases} 3x - 2y = -10 \\ x + 2y = 2 \end{cases}$$

- Karen empties out her purse and finds 45 loose coins, consisting entirely of nickels and pennies. If the total value of the coins is $1.37, how many nickels and how many pennies does she have?

🛠 Building Foundations

Pairs of linear equations form a system of linear equations or _____

The **solution of a system** of linear equations is the set of ordered pairs (or points) that satisfy _____
_____.

To Solve a System of Linear Equations by Graphing

1. Graph both linear equations on _____

2. Observe the point of _____
   a. If the slopes of the two lines are different, then _____
      The system has _____
   b. If the lines have the same slope and different *y*-intercepts, then _____
      The system has _____
   c. If the lines are the same line, then _____
      _____

3. Check the solution (if there is one) in _____

PROCEDURE

9.R.1 Systems of Linear Equations: Solutions by Graphing

> **Consistent and Inconsistent Systems of Linear Equations**
>
> 1. A system is **consistent** if _____
>
> 2. A system is **inconsistent** _____
>
> **DEFINITION**

▶ Watch and Work

Watch the video for Example 3 in the software and follow along in the space provided.

Example 3 Solving Systems (One Solution/A Consistent System)

Solve the system of equations by graphing.

$$\begin{cases} y = -x + 4 \\ y = 2x + 1 \end{cases}$$

Solution

✏ Now You Try It!

Use the space provided to work out the solution to the next example.

Example A Solving Systems (One Solution/A Consistent System)

Solve the system of equations by graphing.

$$\begin{cases} y = 2x + 6 \\ y = -x - 3 \end{cases}$$

> **Dependent and Independent Systems of Linear Equations**
>
> If the graphs of two linear equations are
>
> a. the same line, then the _____
>
> b. different lines, then the _____
>
> **DEFINITION**

9.R.1 Systems of Linear Equations: Solutions by Graphing

👁 Looking Ahead

The following example incorporates the skills you reviewed using graphing to solve a system of linear equations by plotting two equations on the same set of axes and determining their point of intersection. Although graphing systems of equations in two variables is helpful in visualizing the relationship between the equations, there are other techniques that can be used, such as substitution.

Example Preview

Consider the following system of equations.

$$\begin{cases} -5x - 2y = -25 \\ 5x = 5 \end{cases}$$

a. Solve the system by graphing.

b. Solve the system by substitution.

Solution

a. To make it easier for graphing, we will solve the first equation for y, by first writing the equation in slope-intercept form.

$$-5x - 2y = -25$$
$$-2y = 5x - 25$$
$$y = \frac{-5x + 25}{2}$$
$$y = -\frac{5}{2}x + \frac{25}{2}$$

This equation has a y-intercept of $\frac{25}{2} = 12.5$ and a slope of $-\frac{5}{2}$.

The second equation can be simplified as follows.

$$5x = 5$$
$$x = 1$$

The equation $x = 1$ is a vertical line through $(1,0)$ on the x-axis.

Graphing both linear equations on the same set of axes we obtain the following graph.

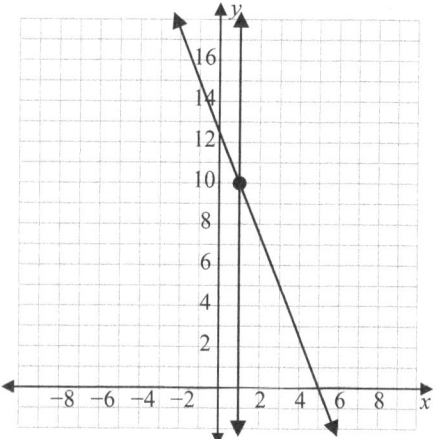

The two lines appear to intersect at the point (1,10). This can be checked by substituting 1 for x and 10 for y in both equations.

b. We can solve either equation for either variable and the choice will not affect the final answer. We will solve for y in the first equation.

$$-5x - 2y = -25$$
$$-2y = 5x - 25$$
$$y = \frac{-5x + 25}{2}$$

We know from the second equation:

$$5x = 5$$
$$x = 1$$

To find the numerical value for y, we substitute 1 for x into the expression we obtained for y in the first step.

$$y = \frac{-5(1) + 25}{2} = 10$$

Therefore, the solution to the system of equations is (1,10).

9.R.1 Exercises

Concept Check

True/False. Determine whether each statement is true or false. If a statement is false, explain how it can be changed so the statement will be true. (**Note:** There may be more than one acceptable change.)

1. To check a solution, substitute it into one of the equations. If the solution satisfies one equation it will satisfy all of the equations.

2. A system of equations with graphs that are parallel lines has exactly one solution.

3. A system of equations with graphs that intersect at one point has exactly one solution.

4. A system of equations with graphs that are the same line has infinitely many solutions.

Practice

Determine which of the given points, if any, lie on both of the lines in the systems of equations by substituting each point into both equations.

5. $\begin{cases} 2x + 4y - 6 = 0 \\ 3x + 6y - 9 = 0 \end{cases}$

   a. $(1, 1)$
   b. $(2, 0)$
   c. $\left(0, \dfrac{3}{2}\right)$
   d. $(-1, 3)$

The graphs of the lines represented by system of equations are given. Determine the solution of the system by looking at the graph. Check your solution by substituting into both equations.

6. $\begin{cases} x + 2y = 4 \\ x - y = -2 \end{cases}$

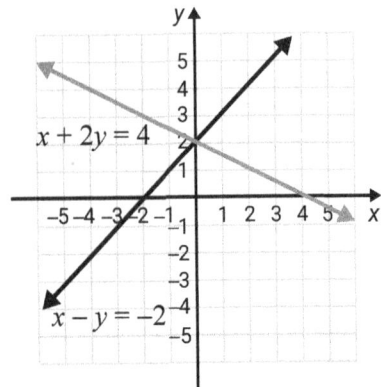

Solve each system of equations by graphing.

7. $\begin{cases} x - 2y = 4 \\ x = 4 \end{cases}$

8. $\begin{cases} 2x + y = 0 \\ 4x + 2y = -8 \end{cases}$

Applications

Each of the following applications has been modeled using a system of equations. Solve the system graphically.

9. **Swimming Pools:** OSHA recommends that swimming pool owners clean their pool decks with a solvent composed of a 12% chlorine solution and a 3% chlorine solution. Fifteen gallons of the solvent consists of 6% chlorine. How much of each of the mixing solutions were used?
Let x = the number of gallons of the 12% solution
and y = the number of gallons of the 3% solution.

The corresponding modeling system is $\begin{cases} x + y = 15 \\ 0.12x + 0.03y = 0.06(15) \end{cases}$

10. ***School Supplies:*** A student bought a calculator and a textbook for a course in algebra. He told his friend that the total cost was $170 (without tax) and that the calculator cost $20 more than twice the cost of the textbook. What was the cost of each item?

    Let x = the cost of the calculator
    and y = the cost of the textbook.

    The corresponding modeling system is $\begin{cases} x + y = 170 \\ x = 2y + 20 \end{cases}$

Writing & Thinking

11. Explain, in your own words, why the answer to a consistent system of linear equations can be written as an ordered pair.

9.R.2 Systems of Linear Equations: Solutions by Substitution

Making Connections

The ability to produce the exact solution to a system of two linear equations in two variables is extremely useful when working on real-world applications. The method of substitution is based on properties of linear equations that you have reviewed previously.

In this section, you will learn skills that you can apply when answering questions like these:

- Solve the following system of equations.

$$\begin{cases} -2x - 6y = -26 \\ -4x - 12y = -49 \end{cases}$$

- How many liters each of a 20% acid solution and a 70% acid solution must be used to produce 50 liters of a 60% acid solution?

- Determine whether the following system of equations is consistent, inconsistent, or dependent.

$$\begin{cases} 6x - 7y = -3 \\ -6x + 5y = 9 \end{cases}$$

Building Foundations

> To Solve a System of Linear Equations by Substitution
> 1. Solve one of the equations for _____
> 2. Substitute the resulting _____
> 3. Solve this new equation, if possible, and then _____
> 4. Check the solution in _____
>
> PROCEDURE

▶ Watch and Work

Watch the video for Example 3 in the software and follow along in the space provided.

Example 3 Solving Systems by Substitution (No Solution)

Use the method of substitution to solve the following system of linear equations.

$$\begin{cases} 3x + y = 1 \\ 6x + 2y = 3 \end{cases}$$

Solution

✏ Now You Try It!

Use the space provided to work out the solution to the next example.

Example A Solving Systems by Substitution (No Solution)

Solve the system:

$$\begin{cases} 2x + y = 1 \\ 10x + 5y = 4 \end{cases}$$

👀 Looking Ahead

As your study of systems of linear equations progresses, you will use the skills you have learned for two variables and apply them to systems of linear equations with three variables, as in the following example.

Example Preview

Solve the following system of equations.

$$\begin{cases} x + 4y = 10 \\ y - 5z = 13 \\ -3x - y + z = 5 \end{cases}$$

Solution

Since two of the equations are already in two variables, we will use the method of substitution to solve this system. First, we will solve for x in equation 1 and substitute this expression into equation 3.

$$x = 10 - 4y \rightarrow -3(10 - 4y) - y + z = 5$$
$$-30 + 12y - y + z = 5$$
$$11y + z = 35$$

Next, we solve for y in equation 2 and substitute this expression into the equation we just found and solve for z.

$$y = 5z + 13 \rightarrow 11(13 + 5z) + z = 35$$
$$143 + 55z + z = 35$$
$$56z = -108$$
$$z = \frac{-27}{14}$$

Now, we can find y from the value we just found for z.

$$y = 13 + 5\left(\frac{-27}{14}\right)$$
$$y = \frac{47}{14}$$

Finally, to find x, we substitute our values for y and z into any of the original equations that contain x.

$$x + 4\left(\frac{47}{14}\right) = 10$$

$$x = \frac{-24}{7}$$

There is only one solution for this system of equations: $\left(\frac{-24}{7}, \frac{47}{14}, \frac{-27}{14}\right)$.

9.R.2 Exercises

Concept Check

True/False. Determine whether each statement is true or false. If a statement is false, explain how it can be changed so the statement will be true. (**Note:** There may be more than one acceptable change.)

1. The method of substitution reduces the problem from one of solving two equations in two variables to solving one equation in one variable.

2. The method of substitution is most often used when one of the equations is impossible to graph.

3. The method of substitution is more accurate than the graphing method.

4. When using the method of substitution, you should always solve the first equation for x.

Practice

Use the method of substitution to solve each system.

5. $\begin{cases} x + y = 6 \\ y = 2x \end{cases}$

6. $\begin{cases} 3x - 7 = y \\ 2y = 6x - 14 \end{cases}$

7. $\begin{cases} 4x = y \\ 4x - y = 7 \end{cases}$

8. $\begin{cases} 3y + 5x = 5 \\ y = 3 - 2x \end{cases}$

Applications

Each of the following applications has been modeled using a system of equations. Use the method of substitution to solve each system.

9. **Rectangles:** The perimeter of a rectangle is 50 meters and the length is 5 meters longer than the width. Find the dimensions of the rectangle.

    Let x = the length and y = the width.

    The corresponding modeling system is $\begin{cases} 2x + 2y = 50 \\ x - y = 5 \end{cases}$

10. **Health & Fitness:** A fitness center manager is trying to decide whether to charge an enrollment fee of $25 with a monthly rate of $50 or an enrollment fee of $100 with a monthly rate of $25. After how many months would it be more profitable for the manager to choose the lower enrollment fee and the higher monthly rate? Round up to the nearest month.

    The corresponding modeling system is $\begin{cases} y = 50x + 25 \\ y = 25x + 100 \end{cases}$

Writing & Thinking

11. Explain the advantages of solving a system of linear equations
    a. by graphing.
    b. by substitution.

9.R.3 Systems of Linear Equations: Solutions by Addition

⟳ Making Connections

Having the ability to solve the same problem using different techniques allows you to select the most efficient approach for a particular question. The method of addition is an alternative method to solve systems of linear equations that avoids possibly cumbersome algebraic steps.

In this section, you will learn skills that you can apply when answering questions like these:

- Solve the following system of equations.

$$\begin{cases} -4x + 3y + 2z = -21 \\ x + 6y - 4z = 9 \\ 5x + 3y - 6z = 30 \end{cases}$$

- Two angles are complementary if the sum of their measures is 90 degrees. Find two complementary angles such that the smaller angle is 194 degrees less than 3 times the larger angle.

- How many ounces of a 14% alcohol solution and a 20% alcohol solution must be combined to obtain 33 ounces of a 16% solution?

🛠 Building Foundations

> **To Solve a System of Linear Equations by Addition**
> 1. Write the equations in _____
> 2. Multiply all terms of one equation by _____
> _____
> 3. Add the two equations by _____
> 4. Back substitute into one of the original equations to _____
> 5. Check the solution (if there is one) in _____
>
> **PROCEDURE**

9.R.3 Systems of Linear Equations: Solutions by Addition

▶ Watch and Work

Watch the video for Example 2 in the software and follow along in the space provided.

Example 2 Solving Systems by Addition (Infinite Solutions)

Use the method of addition to solve the following system of linear equations.

$$\begin{cases} 3x - \dfrac{1}{2}y = 6 \\ 6x - y = 12 \end{cases}$$

Solution

✏ Now You Try It!

Use the space provided to work out the solution to the next example.

Example A Solving Systems by Addition (Infinite Solutions)

Solve the system.

$$\begin{cases} 6x + 3y = 15 \\ 2x + y = 5 \end{cases}$$

Guidelines for Deciding which Method to Use when Solving a System of Linear Equations

1. The graphing method is helpful in "seeing" the _____
   _____

2. Both the substitution method and the _____

3. The substitution method may be reasonable and efficient if _____
   _____

4. The addition method is particularly efficient if _____
   _____

PROCEDURE

👁 Looking Ahead

The method of addition that you have just reviewed in this section is the basis for a more advanced technique, called Gauss-Jordan elimination, which is mainly used for systems with three or more variables. The principle is the same; we want to have coefficients for like terms that are opposites of each other, allowing us to solve for one variable at a time, as you will see in the following example.

Example Preview

Use Gauss-Jordan elimination to solve this system of equations.

$$\begin{cases} x + 3y = -2 \\ y - 5z = 21 \\ -2x - 2y + 2z = 0 \end{cases}$$

Solution

The first thing that needs to be done is to construct the augmented matrix for this system of equations.

$$\begin{bmatrix} 1 & 3 & 0 & | & -2 \\ 0 & 1 & -5 & | & 21 \\ -2 & -2 & 2 & | & 0 \end{bmatrix}$$

Now that we have our starting matrix, the solution for this system of equations can be found in the following manner.

$$\begin{bmatrix} 1 & 3 & 0 & | & -2 \\ 0 & 1 & -5 & | & 21 \\ -2 & -2 & 2 & | & 0 \end{bmatrix} \xrightarrow{2R_1+R_3} \begin{bmatrix} 1 & 3 & 0 & | & -2 \\ 0 & 1 & -5 & | & 21 \\ 0 & 4 & 2 & | & -4 \end{bmatrix} \xrightarrow{-4R_2+R_3} \begin{bmatrix} 1 & 3 & 0 & | & -2 \\ 0 & 1 & -5 & | & 21 \\ 0 & 0 & 22 & | & -88 \end{bmatrix}$$

$$\xrightarrow{\frac{1}{22}R_3} \begin{bmatrix} 1 & 3 & 0 & | & -2 \\ 0 & 1 & -5 & | & 21 \\ 0 & 0 & 1 & | & -4 \end{bmatrix} \xrightarrow{5R_3+R_2} \begin{bmatrix} 1 & 3 & 0 & | & -2 \\ 0 & 1 & 0 & | & 1 \\ 0 & 0 & 1 & | & -4 \end{bmatrix} \xrightarrow{-3R_2+R_1} \begin{bmatrix} 1 & 0 & 0 & | & -5 \\ 0 & 1 & 0 & | & 1 \\ 0 & 0 & 1 & | & -4 \end{bmatrix}$$

After performing all of the necessary elementary row operations to place the augmented matrix for this system in reduced row echelon form, we have the following matrix.

$$\begin{bmatrix} 1 & 0 & 0 & | & -5 \\ 0 & 1 & 0 & | & 1 \\ 0 & 0 & 1 & | & -4 \end{bmatrix}$$

If we now write this matrix in system form, we have the following.

$$\begin{cases} x = -5 \\ y = 1 \\ z = -4 \end{cases}$$

This is equivalent to the original system, but in a form that tells us the solution of the system. Therefore, the ordered triple $(-5, 1, -4)$ solves this system of equations.

9.R.3 Exercises

Concept Check

True/False. Determine whether each statement is true or false. If a statement is false, explain how it can be changed so the statement will be true. (**Note:** There may be more than one acceptable change.)

1. When using the method of addition, the solution only needs to be checked in one of the original equations.

2. It's possible for a system of equations to have no solutions.

3. Both the addition method and the substitution method give approximate solutions.

4. The graphing method is helpful in "seeing" the geometric relationship between the lines and finding approximate solutions.

Practice

Solve each system of linear equations.

5. $\begin{cases} 2x + y = 3 \\ 4x + 2y = 7 \end{cases}$

6. $\begin{cases} y = 2x + 14 \\ x = 14 - 3y \end{cases}$

7. $\begin{cases} 4x - 2y = 8 \\ 2x - y = 4 \end{cases}$

Write an equation for the line determined by the two given points by using the formula $y = mx + b$ to set up a system of equations with m and b as the unknowns.

8. $(2, 3), (1, -2)$

Applications

Each of the following applications has been modeled using a system of equations. Use the method of substitution or the method of addition to solve each system.

9. **Baseball:** A minor league baseball team has a game attendance of 4500 people. Tickets cost $5 for children and $8 for adults. The total revenue made at this game was $26,100. How many adults and how many children attended the game?

    Let x = number of adults
    and y = number of children.

    The system that models the problem is $\begin{cases} x + y = 4500 \\ 8x + 5y = 26{,}100 \end{cases}$

10. **Acid Solutions:** How many liters each of a 30% acid solution and a 40% acid solution must be used to produce 100 liters of a 36% acid solution?

    Let x = amount of 30% solution
    and y = amount of 40% solution.

    The system that models the problem is $\begin{cases} x + y = 100 \\ 0.30x + 0.40y = 0.36(100) \end{cases}$

Writing & Thinking

11. Explain, in your own words, why the answer to a system with infinite solutions is written as an ordered pair with variables.

Name: _____ Date: _____ **311**

9.R.4 Systems of Linear Inequalities

⟳ Making Connections

When solving a linear programming problem in two variables, it is necessary to find the feasible region that satisfies a system of several linear inequalities. Therefore, it is important to review how to graph a system of linear inequalities and to be able to identify the solution set of the system if it exists.

In this section, you will learn skills that you can apply when answering questions like these:

- Construct the constraints and graph the feasible region for the following situation. A plane carrying relief food and water can carry a maximum of 50,000 pounds, and is limited in space to carrying no more than 6000 cubic feet. Each container of water weighs 60 pounds and takes up 1 cubic foot, and each container of food weighs 50 pounds and takes up 10 cubic feet. What is the region of constraint for the number of containers of food and water that the plane can carry?

- Find the minimum and maximum values of the given function $f(x,y) = 6x + 8y$ subject to $x \geq 0$; $y \geq 0$; $4x + y \leq 16$; $x + 3y \leq 15$.

🛠 Building Foundations

To Solve a System of Two Linear Inequalities

1. For each inequality, graph the boundary line and _____

2. Determine the region of the graph that is _____
   _____

   This region is called the _____
   _____

3. To check, pick one test-point in the _____

Note: If there is no intersection, then the system has no solution.

PROCEDURE

▶ Watch and Work

Watch the video for Example 1 in the software and follow along in the space provided.

Example 1 Solving Systems of Linear Inequalities

Solve the system of linear inequalities graphically. $\begin{cases} x \leq 2 \\ y \geq -x + 1 \end{cases}$

Solution

✏ Now You Try It!

Use the space provided to work out the solution to the next example.

Example A Solving Systems of Linear Inequalities

Solve the system of linear inequalities graphically.

$$\begin{cases} y \geq 2 \\ x - y < 4 \end{cases}$$

Possible Solutions to Systems of Linear Inequalities

When the boundary lines are parallel there are three possibilities:

1. The common region will be in the form of _____

2. The common region will be a _____
   _____

3. There will be _____

🔍 Looking Ahead

The following example incorporates the skills you reviewed on how to graph linear inequalities. The constraints of a linear programming problem are expressed by linear inequalities. The graph of these linear inequalities on the same set of axes defines the feasible region. The solution to the linear programming problem is one of the vertices of the feasible region. Substituting the coordinates of the vertices into the objective function provides the solution to the linear programming problem.

Example Preview

On your birthday your rich uncle gave you $19,000. You would like to invest at least $8000 of the money in municipal bonds yielding 4% and no more than $4000 in Treasury bills yielding 5%. How much should be placed in each investment in order to maximize the interest earned in one year? Assume simple interest applies.

Solution

We begin by finding the objective function. Let x represent the amount of money invested in municipal bonds and y represent the amount of money invested in Treasury bills. Applying simple interest for 1 year yields $0.04 \cdot 1 = 0.04$ as the linear coefficient of x. Likewise, the linear coefficient of y is $0.05 \cdot 1 = 0.05$. Thus, the objective function is $f(x,y) = 0.04x + 0.05y$.

Since you have $19,000 to invest, the first constraint is $x + y \leq 19000$. The minimum investment of at least $8000 in municipal bonds translates to $x \geq 8000$. Similarly, the maximum input into Treasury bills can be written as $y \leq 4000$. The final set of constraints, $x \geq 0$ and $y \geq 0$, is logical since the amount of money invested cannot be negative. The portion of the xy-plane that satisfies all four of these inequalities is shown below.

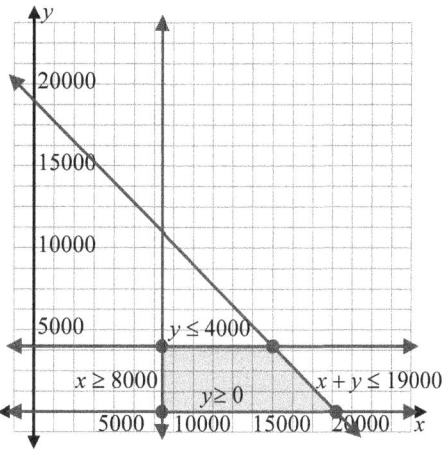

9.R.4 Systems of Linear Inequalities

From here, we can see that the vertices of the feasible region may be represented by the solutions to the following systems of equations.

$$\begin{cases} y = 4000 \\ x+y = 19000 \end{cases}, \begin{cases} y = 0 \\ x+y = 19,000 \end{cases}, \begin{cases} x = 8000 \\ y = 0 \end{cases}, \begin{cases} x = 8000 \\ y = 4000 \end{cases}$$

Solving each of these systems, we find that the vertices are (15,000,4000), (19,000,0), (8000,0), and (8000,4000). We know that one of these vertices must be the point that gives the maximum value. Substituting each of these ordered pairs into the objective function, $f(x,y) = 0.04x + 0.05y$, we find the following.

$$f(15,000,4000) = 800$$
$$f(19,000,0) = 760$$
$$f(8000,0) = 320$$
$$f(8000,4000) = 520$$

We see that income will be maximized by investing $15,000 in municipal bonds and $4000 in Treasury bills.

9.R.4 Exercises

Concept Check

True/False. Determine whether each statement is true or false. If a statement is false, explain how it can be changed so the statement will be true. (**Note:** There may be more than one acceptable change.)

1. When boundary lines are parallel, the system of linear inequalities has no solution.

2. If two half-planes overlap, that region is the union of the graphs.

3. Half-planes are the graphs of linear inequalities.

4. If the graphs of two linear inequalities have no intersection, then the system has no solution.

Practice

Solve the systems of two linear inequalities graphically.

5. $\begin{cases} y > 2 \\ x \geq -3 \end{cases}$

6. $\begin{cases} y > 3x + 1 \\ -3x + y < -1 \end{cases}$

7. $\begin{cases} 2x - 3y \geq 0 \\ 8x - 3y < 36 \end{cases}$

8. $\begin{cases} y > x - 4 \\ y < x + 2 \end{cases}$

Applications

Solve.

9. *Fundraising:* Robin is planning a charity ball to raise money for her favorite charity. There are two different ticket options. The VIP option includes dinner, dancing, and cocktails for $150 per ticket. The regular option includes dancing and cocktails for $75 per ticket. Robin wants to make at least $14,000 in ticket sales. The ballroom that is being used for the charity event has a maximum capacity of 150 people.

   a. Write two linear inequalities to describe the situation. Let the variable x represent the number of VIP tickets sold and let the variable y represent the number of regular tickets sold.

   b. Graph the two linear inequalities on the same coordinate plane.

c. Describe the solution set for the situation.

d. Can Robin reach her sales goal if she only sells tickets for the regular option? Explain why or why not.

Writing & Thinking

10. Graph the inequalities and explain how you can tell that there is no solution.
$$\begin{cases} y \leq 2x - 5 \\ y \geq 2x + 3 \end{cases}$$

Math@Work

Basic Inventory Management 319

Hospitality Management: Preparing for a Dinner Service. 321

Bookkeeper 323

Pediatric Nurse 325

Architecture. 327

Statistician: Quality Control 329

Dental Assistant 331

Financial Advisor 333

Market Research Analyst 335

Chemistry. 337

Astronomy 339

Math Education 341

Physics 343

Forensic Scientist 345

Other Careers in Mathematics 347

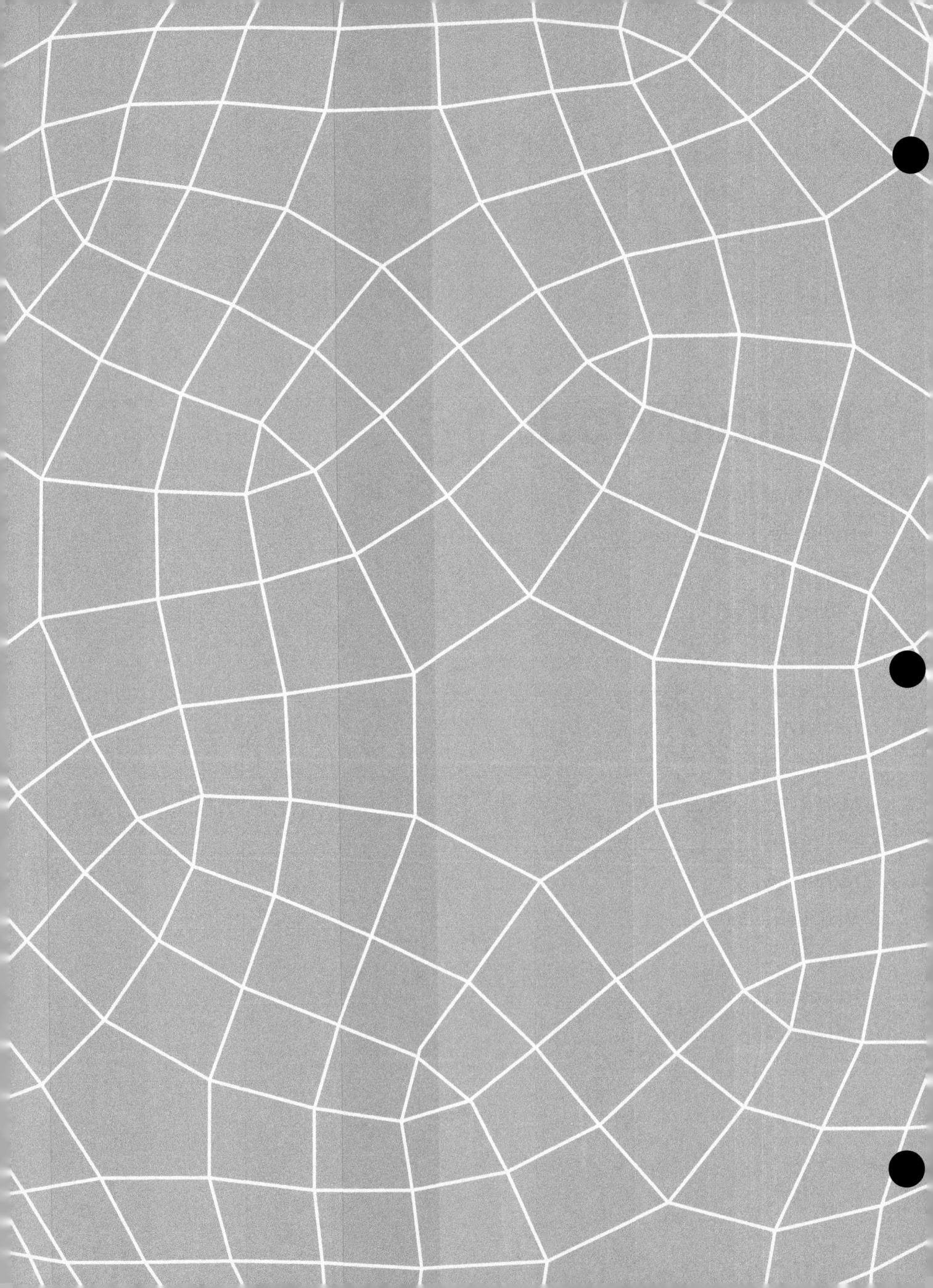

Math@Work

Basic Inventory Management

As a business manager, you will need to evaluate the company's inventory several times per year. While evaluating the inventory, you will need to ensure that enough of each product will be in stock for future sales based on current inventory count, predicted sales, and product cost. Let's say that you check the inventory four times a year, or quarterly. You will be working with several people to get all of the information you need to make the proper decisions. You need the sales team to give you accurate predictions of how much product they expect to sell. You need the warehouse manager to keep an accurate count of how much of each product is currently in stock and how much of that stock has already been sold. You will also have to work with the product manufacturer to determine the cost to produce and ship the product to your company's warehouse. It's your job to look at this information, compare it, and decide what steps to take to make sure you have enough of each product in stock for sales needs. A wrong decision can potentially cost your company a lot of money.

Suppose you get the following reports: an inventory report of unsold products from the warehouse manager and the report on predicted sales for the next quarter (three months) from the sales team.

Unsold Products	
Item	Number in Stock
A	5025
B	150
C	975
D	2000

Predicted Sales	
Item	Expected Sales
A	4500
B	1625
C	1775
D	2150

Suppose the manufacturer gives you the following cost list for the production and shipment of different amounts of each inventory item.

Item	Amount	Cost	Amount	Cost	Amount	Cost
A	500	$875	1000	$1500	1500	$1875
B	500	$1500	1000	$2500	1500	$3375
C	500	$250	1000	$400	1500	$525
D	500	$2500	1000	$4250	1500	$5575

1. Which items and how much of each item do you need to purchase to make sure the inventory will cover the predicted sales?

2. If you purchase the amounts from Problem 1, how much will this cost the company?

3. By ordering the quantities you just calculated, you are ordering the minimum of each item to cover the expected sales. If the actual sales during the quarter are higher than expected, what might happen? How would you handle this situation?

4. Which math skills were necessary to help you make your decisions?

Math@Work

Hospitality Management: Preparing for a Dinner Service

As the manager of a restaurant, you will need to make sure everything is in place for each meal service. This means that you need to predict and prepare for busy times, such as a Friday night dinner rush. To do this, you will need to obtain and analyze information to determine how much of each meal is typically ordered. After you estimate the number of meals that will be sold, you need to communicate to the chefs how much of each item they need to expect to prepare. An additional aspect of the job is to work with the kitchen staff to make sure you have enough ingredients in stock to last throughout the meal service.

You are given the following data, which are the sales records for the signature dishes during the previous four Friday night dinner services.

Week	Meal A	Meal B	Meal C	Meal D
1	30	42	28	20
2	35	38	30	26
3	32	34	26	26
4	30	32	28	22

Meal C is served with a risotto, a type of creamy rice. The chefs use the following recipe, which makes 6 servings of risotto, when they prepare Meal C. (**Note:** The abbreviation for tablespoon is T and the abbreviation for cup is c.)

$5\frac{1}{2}$ c chicken stock $2\frac{1}{3}$ T chopped shallots $\frac{1}{2}$ c red wine

$1\frac{1}{2}$ c rice   2 T chopped parsley $4\frac{3}{4}$ c thinly sliced mushrooms

2 T butter   2 T olive oil $\frac{1}{2}$ c Parmesan cheese

1. For the past four Friday night dinner services, what was the average number of each signature meal served? If the average isn't a whole number, explain why you would round this number either up or down.

2. Based on the average you obtained for Meal C, calculate how much of each ingredient your chefs will need to make the predicted amount of risotto.

3. The head chef reports the following partial inventory: $10\frac{3}{4}$ c rice, $15\frac{3}{4}$ c mushrooms, and 10 T shallots. Do you have enough of these three items in stock to prepare the predicted number of servings of risotto?

4. Which math skills helped you make your decisions?

Math@Work

Bookkeeper

As a bookkeeper, you will often receive bills and receipts for various purchases or expenses from employees of the company you work for. You will need to split the bill by expense code, assign costs according to customer, and reimburse an employee for their out-of-pocket spending. To do this you will need to know the company's reimbursement policies, the expense codes for different spending categories, and which costs fall into a particular expense category.

Suppose two employees from the sales department recently completed sales trips. Employee 1 flew out of state and visited two customers, Customer A and Customer B. This employee had a preapproved business meal with Customer B and was traveling for three days. Employee 2 drove out of state to visit Customer C. This employee stayed at a hotel for the night and then drove back the next day. The expenses for the two employees are as follows.

Employee 1	
Flight and Rental Car	$470.50
Hotel	$278.88
Meals	$110.56
Business Meal	$102.73
Presentation Materials	$54.86

Employee 2	
Miles Driven	578.5 miles
Fuel	$61.35
Hotel	$79.60
Meals	$53.23
Presentation Materials	$67.84

The expense categories used by your company to track spending are: Travel (includes hotel, flights, mileage, etc.), Meals (business), Meals (travel), and Supplies. Traveling employees are reimbursed up to $35 per day for meals while traveling and for all preapproved business meals. They also receive $0.565 per mile driven with their own car, in addition to the amount they spend on fuel.

1. How much will you reimburse each employee for travel meals? Did either employee go over their allowed meal reimbursement amount?

2. What were the total expense amounts reimbursed for each employee?

3. The company you work for keeps track of how much is spent on each customer. When a salesperson visits multiple customers during one trip, the tracked costs are split between the customers. Fill in this table according to how much was spent on each customer for the different expense categories. (**Note:** For meals, only include the amount the employee was reimbursed.)

Expense	Customer A	Customer B	Customer C
Travel			
Meals (business)			
Meals (travel)			
Supplies			
Total			

Name: Date:

Math@Work
Pediatric Nurse

As a pediatric nurse working in a hospital setting, you will be responsible for taking care of several patients during your workday. You will need to administer medications, set IVs, and check each patient's vital signs (such as temperature and blood pressure). While doctors prescribe the medications that nurses need to administer, it is important for nurses to double–check the dosage amounts. Administering the incorrect amount of medication can be detrimental to the patient's health.

During your morning nursing round, you check in on three new male patients and obtain the following information.

	Patient A	Patient B	Patient C
Age	10	9	12
Weight (pounds)	81	68.5	112
Blood Pressure	97/58	100/59	116/73
Temperature (°F)	99.7	97.3	101.4
Medication	A	B	A

The following table shows the bottom of the range for abnormal blood pressure (BP) for boys. If either the numerator or the denominator of the blood pressure ratio is greater than or equal to the values in the chart, this can indicate a stage of hypertension.

Abnormal Blood Pressure for Boys by Age	
	Systolic BP / Diastolic BP
Age 9	109/72
Age 10	111/73
Age 11	113/74
Age 12	115/74

Source: http://www.nhlbi.nih.gov/health/public/heart/hbp/bp_child_pocket/bp_child_pocket.pdf

Medication Directions	
Medication	Dosage Rate
A	40 mg per 10 pounds
B	55 mg per 10 pounds

1. Do any of the patients have a blood pressure that may indicate they have hypertension? If yes, which patient(s)?

2. Use proportions to determine the amount of medication that should be administered to each patient based on weight. Round to the nearest 10 pounds before calculating.

3. The average body temperature is 98.2 degrees Fahrenheit. You are supposed to alert the doctor on duty if any of the patients have a temperature 2.5 degrees higher than average. For which patients would you alert a doctor?

4. Which math skills were necessary to help you make your decisions?

Math@Work

Architecture

As a project architect, you will be part of a team that creates detailed drawings of the project that will be used during the construction phase. It will be your job to ensure that the project will meet guidelines given to you by your company, such as square footage requirements and budget constraints. You will also need to meet the design requirements requested by the client.

Suppose you are part of a team that is designing an apartment building. You are given the task to create the floor plan for an apartment unit with two bedrooms and one bathroom. The apartment management company that has contracted your company to do the project has several requirements for this specific apartment unit.

1. One bedroom is the "master bedroom" and must have at least 60 square feet more than the other bedroom.
2. All walls must intersect or touch at 90 degree angles.
3. The kitchen must have an area of no more than 110 square feet.
4. The apartment must be between 1000 square feet and 1050 square feet.

A preliminary sketch of the apartment is shown here.

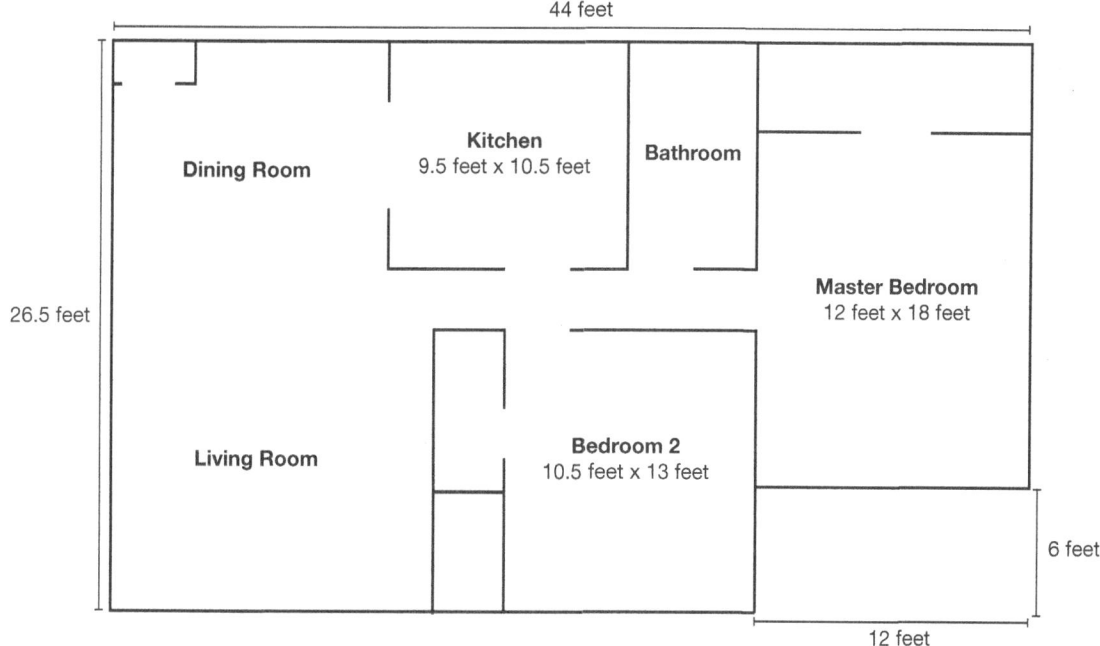

1. Does the apartment have the required total square footage that was requested? Is it over or under the total required?

2. Does the apartment blueprint meet the other requirements given by the client? If not, what does not meet the requirements?

3. For this specific apartment unit, the total construction cost per square foot is estimated to be $75.75. Approximately how much will it cost to construct each two-bedroom apartment based on the floor plan?

Math@Work

Statistician: Quality Control

Suppose you are a statistician working in the quality control department of a company that manufactures the hardware sold in kits to assemble bookshelves, TV stands, and other ready-to-assemble furniture pieces. There are three machines that produce a particular screw and each machine is sampled every hour. A measurement of the screw length is determined with a micrometer, which is a device used to make highly precise measurements. The screw is supposed to be 3 inches in length and can vary from this measurement by no more than 0.1 inches or it will not fit properly into the furniture. The following table shows the screw length measurements (in inches) taken each hour from each machine throughout the day. The screw length data from each machine has also been plotted

Screw Length Measurements (in inches)			
Sample Time	Machine A	Machine B	Machine C
8 a.m.	2.98	2.92	2.99
9 a.m.	3.00	2.94	3.00
10 a.m.	3.02	2.97	3.01
11 a.m.	2.99	2.96	3.03
12 p.m.	3.01	2.94	3.05
1 p.m.	3.00	2.95	3.04
2 p.m.	2.97	2.93	3.06
3 p.m.	2.99	2.92	3.08
Mean			
Range			

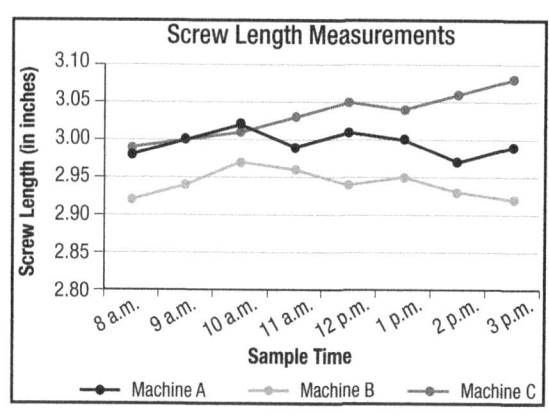

1. Calculate the mean and range of the data for each machine and place them in the bottom two rows of the table.

2. If the screw length can vary from 3 inches by no more than 0.1 inches (plus or minus), what are the lowest and highest values for length that will be acceptable? Place a horizontal line on the graph at each of these values on the vertical axis. These are the tolerance or specification limits for screw length.

3. Have any of the three machines produced an unacceptable part today? Are any of the machines close to making a bad part? If so, which one(s)?

4. Look at the graph and the means from the table that show the average screw length produced by each machine. Draw a bold horizontal line on the graph at 3 to emphasize the target length. Do all the machines appear to be making parts that vary randomly around the target of 3 inches?

5. Look at the range values from the table. Do any of the machines appear to have more variability in the length measurements than the others?

6. In your opinion, which machine is performing best? Would you recommend that any adjustments be made to any of the machines? If so, which one(s) and why?

Math@Work

Dental Assistant

As a dental assistant, your job duties will vary depending on where you work. Suppose you work in a dental office where you assist with dental procedures and managing patients' accounts. When a patient arrives for their appointment, you will need to review their chart and make sure they are up to date on preventive care, such as X-rays and cleanings. When the patient leaves, you will need to fill out an invoice to determine how much to charge the patient for their visit.

Dental patients generally have a new X-ray taken yearly. Cleanings are performed every 6 months, although some patients have their teeth cleaned more often. The following table shows the date of the last X-ray and cleaning for three patients that are visiting the office today. (**Note:** All dates are within the past year.)

Patient Histories		
Patient	**Last X–ray**	**Last Cleaning**
A	April 15	October 20
B	June 6	January 12
C	October 27	October 27

During Patient A's visit, she received a fluoride treatment and a cleaning. Patient A has no dental insurance. During Patient B's visit, he received a filling on one surface of a tooth. Patient B has dental insurance which pays for 60% of the cost of fillings. During Patient C's visit, he had a cleaning, a filling on one surface of a tooth, and a filling on two surfaces of another tooth. Patient C has dental insurance which covers the full cost of cleanings and 50% of the cost of fillings.

Fee Schedule	
Procedure	**Cost**
Cleaning	$95
Fluoride treatment	$35
Filling, One surface	$175
Filling, Two surfaces	$235
X–ray, Panoramic	$110

1. Using today's date, determine which of the three patients are due for a dental cleaning in the next two months?

2. Using today's date, determine which of the patients will require a new set of X-rays during this visit.

3. Determine the amount each patient will be charged for their visit (without insurance). Don't forget to include the cost of any X-rays that are due during the visit.

4. Use the insurance information to determine the amount that each patient will pay out-of-pocket at the end of their visit.

Math@Work
Financial Advisor

As a financial advisor working with a new client, you must first determine how much money your client has to invest. The client may have a lump sum that they have saved or inherited, or they may wish to contribute an amount monthly from their current salary. In the latter case, you must then have the client do a detailed budget so that you can determine a reasonable amount that the client can afford to set aside on a monthly basis for investment.

The second piece of information necessary when dealing with a new client is determining how much risk-tolerance they have. If the client is young or has a lot of money to invest, they may be willing to take more risk and invest in more aggressive, higher interest-earning funds. If the client is older and close to retirement or has little money to invest, they may prefer less-aggressive investments where they are essentially guaranteed a certain rate of return. The range of possible investments that would suit each client's needs and goals are determined using a survey of risk-tolerance.

Suppose you have a client who has a total of $25,000 to invest. You determine that there are two investment funds that meet the client's investment preferences. One option is an aggressive fund that earns an average of 12% interest and the other is a more moderate fund that earns an average of 5% interest. The client desires to earn $2300 this year from these investments.

Investment Type	Principal Invested	·	Interest Rate	=	Interest Earned
Aggressive Fund	x				
Moderate Fund					

To determine the amount of interest earned you know to use the table above and the formula $I = Prt$, where I is the interest earned, P is the principal or amount invested, r is the average rate of return, and t is the length of time invested. Since the initial investment will last one year, $t = 1$.

1. Fill in the Principal Invested and Interest Rate columns of the table with the known information about the principal invested. If x is the amount invested in the aggressive fund and the total amount to be invested is $25,000, create an expression involving x for the amount that will be left to invest in the moderate fund. Place this expression in the appropriate cell of the table.

2. Determine an expression in x for the interest earned on each investment type by multiplying the principal by the interest rate.

3. Determine the amount invested in each fund by setting up an equation using the expressions in the Interest Earned column and the fact that the client desires to earn $2300 from the interest earned on both investments.

4. Verify that the investment amounts calculated for each fund in the previous step are correct by calculating the actual interest earned in a year for each and making sure they sum to $2300.

5. Why would you not advise your client to invest all their money in the fund earning 12% interest, even though it has the highest average interest rate?

Math@Work

Market Research Analyst

As a market research analyst, you may work alone at a computer, collecting and analyzing data, and preparing reports. You may also work as part of a team or work directly with the public to collect information and data. Either way, a market research analyst must have strong math and analytical skills and be very detail-oriented. They must have strong critical-thinking skills to assess large amounts of information and be able to develop a marketing strategy for the company. They must also possess good communication skills in order to interpret their research findings and be able to present their results to clients.

Suppose you work for a shoe manufacturer who wants to produce a new type of lightweight basketball sneaker similar to a product a competitor recently released into the market. You have gathered some sales data on the competitor in order to determine if this venture would be worthwhile, which is shown in the table below. To begin your analysis, you create a scatter plot of the data to see the sales trend. (A scatter plot is a graph made by plotting ordered pairs in a coordinate plane in order to show the relationship between two variables.) You determine that the x-axis will represent the number of weeks after the competitor's new sneaker went on the market and the y-axis will represent the amount of sales in thousands of dollars.

Number of Weeks x	Sales (in 1000s) y
3	15
6	22
9	28
12	35
15	43

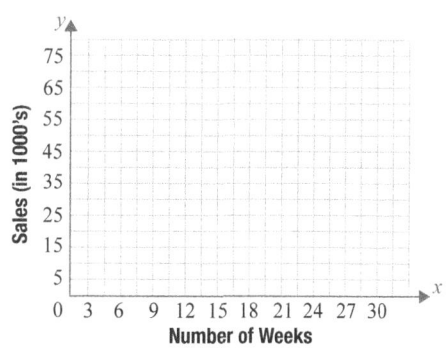

1. Create a scatter plot of the sales data by plotting the ordered pairs in the table on the coordinate plane. Does the data on the graph appear to follow a linear pattern? If so, sketch a line that you feel would "best" fit this set of data. (A market research analyst would typically use computer software to perform a technique called regression analysis to fit a "best" line to this data.)

2. Using the ordered pairs corresponding to weeks 9 and 15, find the equation of a line running through these two data points.

3. Interpret the value calculated for the slope of the equation in Problem 2 as a rate of change in the context of the problem. Write a complete sentence.

4. If you assume that the sales trend in sneaker sales follows the model determined by the linear equation in Problem 2, predict the sneaker sales in 6 months. Use the approximation that 1 month is equal to 4 weeks.

5. Give at least two reasons why the assumption made in Problem 4 may be invalid.

Math@Work

Chemistry

As a pharmaceutical chemist, you will need an advanced degree in pharmaceutical chemistry, which combines biology, biochemistry, and pharmaceuticals. In this career, you will most likely spend your day in a lab setting creating new medications or researching their effectiveness. You will often work as part of a team working towards a joint goal. As a result, in addition to strong math skills and an understanding of chemistry, you will need to have good communication and leadership skills. Since you will be working directly with chemicals, you will also need to have a strong understanding of lab safety rules to ensure the safety of not only yourself but your coworkers as well.

Suppose you work at a pharmaceutical company which creates and produces medications for various skin conditions. You are currently on a team which is developing an acne-controlling facial cleanser. Your team is working on determining the gentlest formula possible that is still effective so that the cleanser can be used on sensitive skin. Half of your team is working with salicylic acid and the other half is working with benzoyl peroxide.

As a part of your work, you will need to keep up on current research. Learning about new chemicals, new methods, and new research will be a continuous part of your life.

1. Perform an Internet search for benzoyl peroxide. How does it work to clean skin and prevent acne?

2. Perform an Internet search for salicylic acid. How does it work to clean skin and prevent acne?

3. Based on your research, which chemical seems better suited to treat acne on sensitive skin?

Another aspect of your career will involve the mixing of chemicals to create new compounds. Having the correct concentrations of chemicals is also important so the resulting solution works as you expect it to. When you don't have the correct concentration of a chemical in stock, it is possible to mix two concentrations together to obtain the desired concentration.

4. Your team wants to create a cleanser with 4% benzoyl peroxide. The lab currently has 2.5% and 10% concentrations of benzoyl peroxide in stock. To create 500 mL of 4% benzoyl peroxide, how much of each concentration should be combined?

Math@Work

Astronomy

Astronomy is the study of celestial bodies, such as planets, asteroids, and stars. While you work in the field of astronomy, you will use knowledge and skills from several other fields, such as mathematics, physics, and chemistry. An important tool of astronomers is the telescope. Several powerful telescopes are housed in observatories around the world. One of the many things astronomers use observatories for is discovering new celestial objects such as a near-Earth object (NEO). NEOs are comets, asteroids, and meteoroids that orbit the sun and cross the orbital path of Earth. The danger presented by NEOs is that they may strike the Earth and result in global catastrophic damage. (**Note:** The National Aeronautics and Space Administration (NASA) keeps track of all NEOs which are a potential threat at the website http://neo.jpl.nasa.gov/risk/)

For an asteroid to be classified as an NEO, the asteroid must have an orbit that partially lies within 0.983 and 1.3 astronomical units (AU) from the sun, where 1 AU is the furthest distance from the Earth to the sun, approximately 9.3×10^7 miles.

Near-Earth Object Distance			
	Minimum		**Maximum**
Distance in AU	0.983 AU	1 AU	1.3 AU
Distance in Miles		9.3×10^7 miles	

Suppose you discover three asteroids that you suspect may be NEOs. You perform some calculations and come up with the following facts. The furthest that Asteroid A is ever from the sun is 81,958,000 miles. The closest Asteroid B is ever to the sun is 125,290,000 miles. The closest Asteroid C is ever to the sun is 92,595,000 miles.

1. To determine if any of the asteroids pass within the range to be classified as an NEO, fill in the missing values from the table.

2. Based on the measurements from Problem 1, do any of the three asteroids qualify as an NEO?

There are two scales that astronomers use to explain the potential danger of NEOs. The Torino Scale is a scale from 0 to 10 that indicates the chance that an object will collide with the Earth. A rating of 0 means there is an extremely small chance of a collision and a 10 indicates that a collision is certain to happen. The Palermo Technical Impact Hazard Scale is used to rate the potential impact hazard of an NEO. If the rating is less than −2, the object poses a very minor threat with no drastic consequences if the object hits the Earth. If the rating is between −2 and 0, then the object should be closely monitored as it could cause serious damage.

Go to the NASA website http://neo.jpl.nasa.gov/risk/ to answer the following questions.

3. Does any NEO have a Torino Scale rating higher than 0? If so, what is the object's designation (or name) and during which year range could a potential impact occur?

4. Which NEO has the highest Palermo Scale rating? During which year range could a potential impact occur?

Math@Work

Math Education

As a math instructor at a public high school, your day will be spent preparing class lectures, grading assignments and tests, and teaching students with a wide variety of backgrounds. While teaching math, it is your job to explain the concepts and skills of math in a variety of ways to help students learn and understand the material. As a result, a solid understanding of math and strong communication skills are very important. Teaching math is a challenge and being able to understand the reasons that students struggle with math and empathize with these students is a critical aspect of the job.

Suppose that the next topics you plan to teach to your algebra students involve finding the greatest common factor and factoring by grouping. To teach these skills, you will need to plan how much material to cover each day, choose examples to walk through during the lecture, and assign in-class work and homework. You decide to spend the first day on this topic explaining how to find the greatest common factor of a list of integers.

1. It is usually easier to teach a group of students a new topic by initially showing them a single method. If a student has difficulty with that method, then showing the student an alternative method can be helpful. Which method for finding the greatest common factor would you teach to the class during the class lecture?

2. On a separate piece of paper, sketch out a short lecture on finding the greatest common factor of a list of integers. Be sure to include examples that range from easy to difficult.

3. While the class is working on an in-class assignment, you find that a student is having trouble following the method that you taught to the entire class. Describe an alternative method that you could show the student.

4. From your experience with learning how to find the greatest common factor of a list of integers, what do you think are some areas that might confuse students and cause them to struggle while learning this topic? Explain how understanding the areas that might cause confusion can help you become a better teacher.

Math@Work

Physics

As an employee of a company that creates circuit boards, your job may vary from designing new circuit boards, setting up machines to mass–produce the circuit boards, to testing the finished circuit boards as part of quality control. Depending on your position, you may work alone or as part of a team. Regardless of who you work with, you will need strong math skills to be able to create new circuit board designs and strong communication skills to describe the specifications for a new circuit board design, describe how to set up the production line, or explain why a part is faulty.

Suppose your job requires you to create new circuit boards for a variety of electronic equipment. The latest circuit board that you are designing is a small part of a complicated device. The circuit board you create has three resistors which run in parallel, as shown in the diagram.

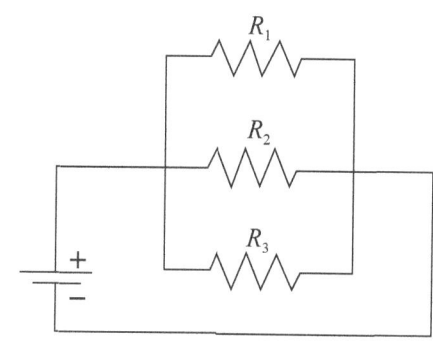

Two of the resistors were properly labeled with their correct resistance, which is measured in ohms. The first resistor has a rating of 2 ohms. The second resistor has a rating of 3 ohms. The third resistor was taken from the supply shelf for resistors of a certain rating, but the resistor was unlabeled. As a result, you are unsure if it has the correct resistance for the current you want to produce. You use an ohmmeter, a device that measures resistance in a circuit, to determine that the total resistance of the circuit you created is $\frac{30}{31}$ ohms.

You know that the equation to determine the total resistance R_t is $\frac{1}{R_t} = \frac{1}{R_1} + \frac{1}{R_2} + \frac{1}{R_3}$, where R_1 is the resistance of the first resistor, R_2 is the resistance of the second resistor, and R_3 is the resistance of the third resistor.

1. Use the formula to determine the resistance of the third resistor given that the total resistance of the circuit is $\frac{30}{31}$ ohms.

2. Was the third resistor on the correct shelf if you took it from the supply shelf that holds resistors with a rating of 7 ohms?

3. What would be the total resistance of the circuit if the third resistor had a rating of 7 ohms?

4. What do you think would happen if the resistance of the unlabeled resistor wasn't determined and the circuit board was sent to the production line to be mass–produced?

Math@Work

Forensic Scientist

As a forensic scientist, you will work as part of a team to investigate the evidence from a crime scene. Every case you encounter will be unique and the work may be intense. Communication is especially important because you will need to be clear and honest about your findings and your conclusions. A suspect's freedom may depend on the conclusions your team draws from the evidence.

Suppose the most recent case that you are involved in is a hit-and-run accident. A body was found at the side of the road with skid marks nearby. The police are unsure if the cause of death of the victim was vehicular homicide. Among the case description, the following information is provided to you.

Accident Report	
Date:	June 14
Time:	9:30 pm
Climate:	55 degrees Fahrenheit, partly cloudy, dry
Description of crime scene:	
Victim was found at the side of a road. Body temperature upon arrival is 84.9 °F. Posted speed limit is 30 mph. Road is concrete. Conditions are dry. Skid marks near the body are 88 feet in length.	

Known formulas and data:

A body will cool at a rate of 2.7 °F per hour until the body temperature matches the temperature of the environment. Average human body temperature is 98.6 °F.

Impact Speed and Risk of Death	
Impact Speed	Risk of Death
23 mph	10%
32 mph	25%
42 mph	50%
58 mph	90%
Source: 2011 AAA Foundation for Traffic Safety "Impact Speed and Pedestrian's Risk of Severe Injury or Death"	

Braking distance is calculated using the formula $\frac{s}{\sqrt{l}} = k$, where s is the initial speed of the vehicle in mph, l is the length of the skid marks in feet, and k is a constant that depends on driving conditions. Based on the driving conditions on that road for the last 12 hours, $k = \sqrt{20}$.

1. Based on the length of the skid marks, how fast was the car traveling before it attempted to stop? Round to the nearest whole number.

2. Based on the table, what percent of pedestrians die after being hit by a car moving at that speed?

3. Based on the cooling of the body, if the victim died instantly, how long ago did the accident occur? Round to the nearest hour.

4. Can you think of any other factors that should be taken into consideration before determining whether the impact of the car was the cause of death?

Name: Date:

Math@Work

Other Careers in Mathematics

Earning a degree in mathematics or minoring in mathematics can open many career pathways. While a degree in mathematics or a field which uses a lot of mathematics may seem like a difficult path, it is something anyone can achieve with practice, patience, and persistence. Three growing fields of study which rely on mathematics are actuarial science, computer science, and operations research. While each of these fields involves mathematics, they require special training or additional education outside of a math degree. A brief description of each career is provided below along with a source to find more information about these careers.

Growing Fields of Study

Actuarial Science: The field of actuarial science uses methods of mathematics and statistics to evaluate risk in industries such as finance and insurance. Visit www.beanactuary.org for more information

Computer Science: From creating web pages and computer programs to designing artificial intelligence, computer science uses a variety of mathematics. Visit www.acm.org for more information.

Operations Research: The discipline of operations research uses techniques from mathematical modeling, statistical analysis, and mathematical optimization to make better decisions, such as maximizing revenue or minimizing costs for a business. Visit www.informs.org for more information.

There are numerous careers that have not been discussed in this workbook. Exploring career options before choosing a major is a very important step in your academic career. Learning about the career you are interested in before completing your degree can help you choose courses that will align with your career goals. You should also explore the availability of jobs in your chosen career and whether you will have to relocate to another area to be hired. The following web sites will help you find information related to different careers that use mathematics. Another great resource is the mathematics department at your college.

The **Mathematical Association of America** has a website with information about several careers in mathematics. Visit www.maa.org/careers to learn more.

The **Society for Industrial and Applied Mathematics** also has a webpage dedicated to careers in mathematics. Visit www.siam.org/careers to learn more.

The **Occupational Outlook Handbook** is a good source for information on educational requirements, salary ranges, and employability of many careers, not just those that involve mathematics. Visit http://www.bls.gov/ooh/ to learn more.

Answer Key

Chapter 1.R: Algebraic Expressions, Equations, and Inequalities

1.R.1 Exercises

Concept Check

1. False; Equals 81
3. False; 7^0 is 1.
5. True
7. False; 7605 is divisible by 5.
9. False; A prime number has exactly 2 factors.
11. False; 231 is a composite number.
13. False; The LCM of 15 and 25 is 75.
15. True

Practice

17. a. 4 b. 0 c. 1
19. 3, 5
21. None
23. Prime
25. 5^3
27. 24
29. 15

Applications

31. a. No. The formula incorrectly shows that each shirt is $10, and the total cost would be $25 − $10 · 11 = $25 − $110 = −$85.
    b. $165; ($25 − $10)·11
33. 4 team members would work 110 hours each; 8 team members would work 55 hours each.
35. 1, 2, 3, 4, 6, 8, 9, 12, 18, 24, 36, 72

Writing & Thinking

37. If addition is within parentheses (or other grouping symbols), addition would be performed first.
39. No, some odd numbers are the product of two or more odd prime factors, for example, 3 · 3 = 9, 3 · 5 = 15, 3 · 7 = 21, etc.
41. Since the LCM is constructed using the prime factors of each number in the set, by definition, each number will divide the LCM.

1.R.2 Exercises

Concept Check

1. True
3. False; The statement $\frac{1}{3} \cdot \frac{2}{5} = \frac{2}{5} \cdot \frac{1}{3}$ is an example of the commutative property of multiplication.
5. False; The reciprocal of 1 is 1.
7. False; The reciprocal of 12 is $\frac{1}{12}$.

Practice

9. $\frac{1}{4}$
11. $\frac{1}{3}$
13. $\frac{8}{9}$
15. Undefined

Applications

17. $\frac{3}{8}$
19. 200 years
21. No. If a fraction is less than 1 then its product with another number will be less than that other number. So, if the other number is less than 1, the product will be less than 1. Answers will vary.
23. $0 = \frac{0}{1}$ and the reciprocal would be $\frac{1}{0}$ but division by 0 is undefined. So 0 has no reciprocal.

1.R.3 Exercises

Concept Check

1. True
3. False; LCD stands for least common denominator.
5. True

Practice

7. $\frac{17}{21}$
9. $\frac{11}{15}$
11. $\frac{23}{42}$
13. $-\frac{1}{4}$

Applications

15. 1 ounce

Writing & Thinking

17. The LCM finds the least common multiple of a set of numbers. The LCD does the same thing for the set of numbers determined by the denominators.

1.R.4 Exercises

Concept Check

1. False; A proportion is a statement that two ratios are equal.
3. True
5. True

Practice

7. False
9. $B = 7.8$

Applications

11. 180 minutes or 3 hours

1.R.5 Exercises

Concept Check

1. True
3. False; A decimal number that is between 0.01 and 0.10 is between 1% and 10%.
5. False; Fractions that have denominators other than 100 can be changed to a percent.

Practice

7. 20%
9. 2%
11. 0.07
13. 75%
15. $1\frac{1}{5}$

Applications

17. 4%
19. 85%

Writing & Thinking

21. Percent means per centum or per 100. For example, fifty-eight percent means 58 out of 100. Percent can be written as a fraction with 100 in the denominator as in 58/100. The decimal equivalent, 0.58 is read as "fifty-eight hundredths," indicating percent can be written using the hundredths place, another connection.

23. 100% = 1 so anytime there is a mixed number, which has a value greater than 1, the percentage will be greater than 100%. Proper fractions (numerator is smaller than denominator) have a value less than 1 and therefore the percentage will be less than 100%.

Chapter 2.R: Equations and Inequalities in One Variable

2.R.1 Exercises

Concept Check

1. True
3. True

Practice

5. [number line with points at −3, −2, 0, 1]
7. 0, 4, 8
9. True

Applications

11. −4500 meters

Writing & Thinking

13. If y is a negative number then $-y$ represents a positive number. For example, if $y = -2$, then $-y = -(-2) = 2$.

2.R.2 Exercises

Concept Check

1. False; The sum of a positive and negative number can be positive, negative, or zero.
3. False; The sum of two positive numbers is always positive, zero is neither positive nor negative.

Practice

5. −6
7. −0.5

Applications

9. **a.** $45,000 + (−$8000) + (−$2000) + $15,000.
b. $50,000.

Writing & Thinking

11. $|0| + |0| = 0$

2.R.3 Exercises

Concept Check

1. False; The sum of a number and its additive inverse is zero.
3. True

Practice

5. −11
7. 3
9. −15

Applications

11. −18°F (a decrease of 18 degrees Fahrenheit)

Writing & Thinking

13. Add the opposite of the second number to the first number.

2.R.4 Exercises

Concept Check

1. True.
3. False; The product and quotient will be positive.

Practice

5. 48
7. 2

Applications

9. −12

Writing & Thinking

11. Negative; The product of every two negative numbers will be positive and this result multiplied by the remaining negative will give a negative answer.

Chapter 3.R: Equations and Inequalities in Two Variables

3.R.1 Exercises

Concept Check

1. **a.** True
   b. False; Not all rectangles have four equal sides.
3. True
5. True
7. False; The length of the radius of a circle is half of the length of the diameter.
9. True
11. **a.** B
   b. C
   c. A
   d. E
   e. D

Practice

13. 44 cm
15. 45 cm
17. 50 in.
19. 70 in.^{3}
21. 81 ft^2
23. 48 in.^{2}
25. 36 cm^2
27. **a.** 35.98 in.
   b. 76.93 in.^{2}
29. 7536 m^2
31. 226.08 m^2

Applications

33. 135 square feet
35. **a.** 63.6 square inches
   b. $0.13 per square inch

Writing & Thinking

37. To find the volume of the ice cream itself, the volume is half the volume of a sphere and the formula used would be $V = \dfrac{1}{2} \cdot \dfrac{4}{3}\pi r^3 = \dfrac{2}{3}\pi r^3$. The result is the volume of the ice cream that is exposed. Next find the volume of the cone by using the formula $V = \dfrac{1}{3}\pi r^2 h$. The volume of the ice cream cone will be the sum of these two volumes.

39. One diameter is equal to two radii. Thus, $d = 2r$ and $C = 2\pi r = \pi d$.

3.R.2 Exercises

Concept Check

1. True
3. True

Practice

5. 6
7. Yes; $6^2 + 8^2 = 10^2$
9. $c = 5$

Applications

11. 17.0 inches

3.R.3 Exercises

Concept Check

1. False; If the original number is negative, the principal square root will not be the same as the original number.
3. False; The radicand is underneath the radical sign.

Practice

5. 7
7. 10
9. 0.2

Applications

11. a. 4 cm
    b. 20 cm

Writing & Thinking

13. Cubing a negative real number is equivalent to multiplying a negative number by itself 3 times. The product of three negative numbers is negative.

3.R.4 Exercises

Concept Check

1. True
3. False; If x is a real number, then $\sqrt{x^2} = |x|$

Practice

5. $9\sqrt{2}$
7. $2x^5y\sqrt{6x}$
9. $-2x^2\sqrt[3]{x^2}$

Applications

11. $\sqrt{3} \approx 1.73$ amperes

Writing & Thinking

13. A cube root has no restrictions as the cube root of a negative number is negative.

3.R.5 Exercises

Concept Check

1. True
3. True

Practice

5. $\begin{cases} A(-5,1), B(-3,3), \\ C(-1,1), D(1,2), \\ E(2,-2) \end{cases}$

7. a. $(0,-1)$
   b. $(4,1)$
   c. $(2,0)$
   d. $(8,3)$

9. b, c

Applications

11. a.

D	E
100	85
200	170
300	255
400	340
500	425

b.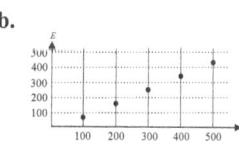

3.R.6 Exercises

Concept Check

1. False; The addition and multiplication principles of equality can be used with decimal or fractional coefficients.
3. True

Practice

5. $x = -3$
7. $x = -\dfrac{27}{10}$

Applications

9. 14,000 tickets per hour

Writing & Thinking

11. a. The 4 should have been multiplied by 3 so that the 3 was distributed over the entire left-hand side of the equation; Correct answer is $x = 15$.
    b. 3 should be subtracted from each side, not from each term, and $5x - 3$ doesn't simplify to $2x$; Correct answer is $x = \dfrac{8}{5}$.

3.R.7 Exercises

Concept Check

1. True
3. False; It is called a contradiction.

Practice

5. $x = -5$
7. $x = -1$
9. Contradiction

Applications

11. 20 guests

Writing & Thinking

13. a. $5x + 1$
    b. $x = 6$
    c. Answers will vary.

3.R.8 Exercises

Concept Check

1. True
3. False; Only one value in the solution set needs to be checked.

Practice

5. ←————|————→
         -3
   Half-open interval

7. $(4, \infty)$
   ←————|————→
        0 4

9. $[4, \infty)$
   ←————|————→
          4

Applications

11. a. The student must score at least 102 points, which is not possible. Thus he cannot earn an A in the course.
    b. The student must score at least 192 points to earn an A in the course.

Writing & Thinking

13. a. Answers will vary.
    b. Answers will vary.

3.R.9 Exercises

Concept Check

1. False; Radical equations may also have one or no solution.
3. False; The process needs to be repeated until all radicals have been eliminated.

Practice

5. $x = 3$
7. $x = 4$

Applications

9. 12 ft

Chapter 4.R: Relations, Functions, and Their Graphs

4.R.1 Exercises

Concept Check

1. True
3. True

Practice

5. $\{(-5,-4),(-4,-2),(-2,-2),(1,-2),(2,1)\}$;
   $D = \{-5,-4,-2,1,2\}$;
   $R = \{-4,-2,1\}$;
   Function

7. Not a function;
   $D = (-\infty, \infty)$;
   $R = (-\infty, \infty)$

9. **a.** -10 **b.** 86 **c.** 86

4.R.2 Exercises

Concept Check

1. True
3. False; Subtraction is indicated by the phrase "five less than a number."

Practice

5. $x + 6$
7. $\dfrac{x}{2} - 18$
9. **a.** $4n - 6$ **b.** $6 - 4n$
11. The product of a number and negative nine

Writing & Thinking

13. The Commutative Property of Addition and Multiplication permits the order of items being added or multiplied to change and still have the same result. This property does not hold true for subtraction or division. Therefore, order is important for subtraction and division problems or the answer will change or be incorrect.

4.R.3 Exercises

Concept Check

1. True
3. False; Odd integers are integers that are not even.

Practice

5. $x - 5 = 13 - x$; 9
7. $n + (n+1) + (n+2) = 93$; $30, 31, 32$

Applications

9. **a.** The unknown value is the length of the call in minutes.
   b. $m = 20$
   c. The collect call lasted 20 minutes.

Writing & Thinking

11. **a.** $n, n+2, n+4, n+6$
    b. $n, n+2, n+4, n+6$
    c. Yes; Answers will vary.

4.R.4 Exercises

Concept Check

1. False; Variables need to be considered as well.
3. True

Practice

5. 5
7. $7(2x + 3)$
9. $(3 + a)(x + y)$

Applications

11. **a.** 32 feet
    b. $16x(3 - x)$
    c. 32 feet
    d. Yes; They are equivalent expressions.

4.R.5 Exercises

Concept Check

1. True
3. True

Practice

5. $(x - 9)(x + 3)$
7. $(y - 12)(y - 2)$

Applications

9. Base $= x + 48$; Height $= x$

Writing & Thinking

11. If the sign of the constant term is positive, the signs in the factors will both be positive or both be negative. If the sign of the constant term is negative, the sign in one factor will be positive and the sign in the other factor will be negative.

4.R.6 Exercises

Concept Check

1. False; The middle term should be the sum of the inner and outer products.
3. False; The first step is to multiply a and c.

Practice

5. $(6x + 5)(x + 1)$
7. Not factorable
9. $2(2x - 5)(3x - 2)$

Writing & Thinking

11. This is not an error, but the trinomial is not completely factored. The completely factored form of this trinomial is $2(x + 2)(x + 3)$.

4.R.7 Exercises

Concept Check

1. False; The first step is to check for a common monomial factor.
3. False; It might be factorable by the grouping method.

Practice

5. $(x + 5)^2$
7. $(5x + 6)(4x - 9)$
9. $(x + 5)(x^2 - 5x + 25)$
11. $(x + 12)(x - 3)$
13. $2(2x - 1)(x - 3)$

4.R.8 Exercises

Concept Check

1. True
3. True

Practice

5. $x = 2, 9$
7. $x = -4, 6$

Applications

9. **a.** 640 ft; 384 ft
   b. 144 ft; 400 ft
   c. 7 seconds;
   $0 = -16(t + 7)(t - 7)$

Writing & Thinking

11. This allows for use of the zero factor property which says that for the product to equal zero one of the factors must equal zero. Answers will vary.

4.R.9 Exercises

Concept Check

1. False; i, $-i$, 1, and -1

3. False; The product is -1.

Practice

5. $-28 - 24i$

7. $0 - \dfrac{5}{4}i$

9. $0 + i$

Writing & Thinking

11. Given a complex number $(a+bi)$: $(a+bi)(a-bi)$
$= a^2 - abi + abi - b^2 i^2$
$= a^2 + b^2$
which is the sum of squares of real numbers. Thus the product must be a positive real number.

4.R.10 Exercises

Concept Check

1. True

3. False; Two real solutions

Practice

5. 68; Two real solutions

7. $x = -2 \pm 2\sqrt{2}$

9. $x = 1, \dfrac{4}{3}$

Writing & Thinking

11. $x^4 - 13x^2 + 36 = 0$; multiplied
$(x-2)(x+2)(x-3)(x+3)$

Chapter 5.R: Working with Functions

5.R.1 Exercises

Concept Check

1. True.
3. True.

Practice

5. a. 36
   b. 16
7. -10
9. 129
11. 143

Applications

13. a. $-\$42 - \$35 - (3 \cdot \$5)$.
    b. $-\$92$.

Writing & Thinking

15. Smaller; When any positive number is multiplied by a fraction (or decimal) between 0 and 1, the result will be smaller. This is what is happening when a number between 0 and 1 is squared. Answers will vary.

5.R.2 Exercises

Concept Check

1. True.
3. False; In the term "12a," 12 is the coefficient.

Practice

5. -5, 3, and 8 are like terms; $7x$ and $9x$ are like terms.
7. $10x$
9. $3y + 4$; 13

Applications

11. $50,000

Writing & Thinking

13. Like terms have the same variables with the same exponents. For example, $4a^2bc^3$ and $-3a^2bc^3$ are like terms. Unlike terms either have different variables or possibly the same variables with different exponents. For example, $6ab$ and $-9a^2b$ are unlike terms and $5xy$ and $13ax$ are unlike terms.

5.R.3 Exercises

Concept Check

1. False; The distributive property can be used when multiplying any types of polynomials.
3. True

Practice

5. $-4x^8 + 8x^7 - 12x^4$
7. $y^3 + 2y^2 + y + 12$

Applications

9. $V = 10x^2 + 220x + 1200$ cubic inches

5.R.4 Exercises

Concept Check

1. True
3. False; Missing powers should be filled in with zeros.

Practice

5. $y^2 - 2y + 3$
7. $x - 6 + \dfrac{4}{x+4}$
9. $x^3 + 2x + 5 + \dfrac{17}{x-3}$

Applications

11. a. $x^2 - 4x - 5$ square inches
    b. $x^2 - 3x - 10$ square inches

5.R.5 Exercises

Concept Check

1. True
3. False; A rational expression cannot have a zero denominator.

Practice

5. $\dfrac{3x}{4y}$; $x \neq 0, y \neq 0$
7. $\dfrac{x-3}{y-2}$; $y \neq -2, 2$
9. $\dfrac{1}{x-3}$; $x \neq 0, 3$

Applications

11. a. $p(x) = \dfrac{15x + 200}{x}$
    b. $35
    c. $x \neq 0$
    d. The variable cannot be negative because you cannot have a negative quantity of people. There would also be a maximum number depending on the size of the room.

Writing & Thinking

13. a. A rational expression is an algebraic expression that can be written in the form $\dfrac{P}{Q}$ where P and Q are polynomials and $Q \neq 0$.
    b. $\dfrac{x-1}{(x+2)(x-3)}$; Answers will vary.

c. $\frac{1}{x+5}$;
   Answers will vary.

5.R.6 Exercises

Concept Check

1. False; The reciprocal is $\frac{x+3}{x}$.
3. False; the restriction is 0.

Practice

5. $\frac{x+3}{x}$

7. $\frac{x}{12}$

Applications

9. a. $\frac{x^2 - 3x - 10}{x+3}$
   b. $\frac{x^2 + 5x + 6}{x-5}$
   c. $(x+2)^2 = x^2 + 4x + 4$

5.R.7 Exercises

Concept Check

1. True
3. True

Practice

5. $\frac{4}{5xy}$

7. $\frac{11}{2(1+4x)}$

9. $\frac{-5}{x+1}$

Applications

11. a. $\frac{4r + r^2}{4}$

b. 0.0609
c. 0.0609
d. Yes; They are calculated from different forms of the same expression.
e. 0.0009
f. Answers will vary. It is larger because the interest is compounded along with the principal.

Chapter 7.R: Exponential and Logarithmic Functions

7.R.1 Exercises

Concept Check

1. False; If there is no exponent written, the exponent is assumed to be 1.
3. True

Practice

5. y^{11}
7. $\frac{1}{x}$
9. $-18x^5 y^7$

Applications

11. 2^8 GB

7.R.2 Exercises

Concept Check

1. True
3. False; The rules for exponents can be applied in any order, resulting in the same answer.

Practice

5. 64
7. $-\frac{2y^6}{27x^{15}}$
9. $\frac{y^8}{16x^8}$

7.R.3 Exercises

Concept Check

1. True
3. True

Practice

5. $\sqrt[3]{8}$
7. $\frac{1}{10}$
9. $a^{\frac{1}{4}}$

Applications

11. 576 ft²

Writing & Thinking

13. No; $\sqrt[5]{a} \cdot \sqrt{a} = a^{\frac{7}{10}}, \sqrt[5]{a^2} = a^{\frac{2}{5}}$, $a^{\frac{7}{10}} \neq a^{\frac{2}{5}}$

7.R.4 Exercises

Concept Check

1. True
3. True

Practice

5. $\log_7 49 = 2$
7. $x = -3$
9. $x = 3.7$

Writing & Thinking

11. The two functions are symmetric about the x-axis.

Chapter 8.R: Conic Sections

8.R.1 Exercises

Concept Check

1. False; The product will be a binomial.
3. True

Practice

5. $x^2 - 14x + 49$; Perfect square trinomial
7. $x^2 + 8x + 16$; Perfect square trinomial
9. $9x^2 - 12x + 4$; Perfect square trinomial

Applications

11. a. $A(x) = 400 - 4x^2$
    b. $P(x) = 4(20 - 2x) + 8x$
       $= 80$

Writing & Thinking

13. $(x+5)^2 = x^2 + 2(5x) + 5^2$. Answers will vary.

8.R.2 Exercises

Concept Check

1. True
3. False; The sum of two squares is not factorable.

Practice

5. $(y-8)^2$

7. $2(x-8)(x+8)$

9. $(3x-y)(3x+y)$

Writing & Thinking

11. a. $xy + xy + x^2 + y^2$
$= x^2 + 2xy + y^2$
$= (x+y)^2$

b. $(x+y)(x+y)$
$= (x+y)^2$

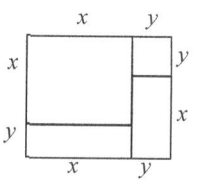

Chapter 9.R: Systems of Equations and Inequalities

9.R.1 Exercises

Concept Check

1. False; The solution must be checked in all equations.
3. True

Practice

5. a, c
7. (4, 0)

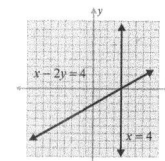

Applications

9.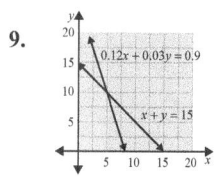

5 gallons of 12%; 10 gallons of 3%

Writing & Thinking

11. The solution to a consistent system of linear equations is a single point, which is easily written as an ordered pair.

9.R.2 Exercises

Concept Check

1. True
3. True

Practice

5. (2, 4)
7. No solution

Applications

9. 15 m × 10 m

Writing & Thinking

11. Answers will vary.

9.R.3 Exercises

Concept Check

1. False; The solution always needs to be checked in both original equations.
3. False; Both methods give exact solutions.

Practice

5. No solution
7. $(x, 2x - 4)$

Applications

9. 1200 adults; 3300 children

Writing & Thinking

11. Answers will vary.

9.R.4 Exercises

Concept Check

1. False; When boundary lines are parallel, the solution is either the strip between the boundary lines, a half-plane, or there is no solution.
3. True

Practice

5.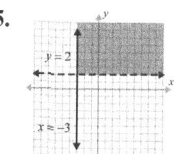

7.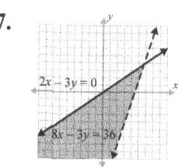

Applications

9. a. $\begin{cases} 150x + 75y \geq 14{,}000 \\ x + y \leq 150 \end{cases}$

b.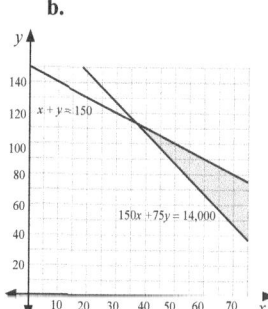

c. Answers will vary.

d. No. $150 \cdot \$75 = \$11{,}250$

Notes

Notes

Notes